Artificial Intelligence Applications For Business

Computer-Based Information Systems in Organizations

edited by
Michael J. Ginzberg
New York University

Artificial Intelligence Applications for Business • *Walter Reitman, Editor*

In Preparation:
The MIS Context: Social and Political Forces in the Information Systems Environment • *Michael Ginzberg*

Artificial Intelligence Applications For Business

Proceedings of the NYU Symposium, May, 1983

edited by
Walter Reitman
BBN Laboratories

ABLEX PUBLISHING CORPORATION
NORWOOD, NEW JERSEY 07648

Printed in the United States of America.

Library of Congress Cataloging in Publication Data
Main entry under title:

Artificial intelligence applications for business.

   (Computer-based information systems in organizations)
   Includes indexes.
   1. Business—Data processing—Addresses, essays,
lectures.   2. Artificial intelligence—Addresses, essays,
lectures.   I. Reitman, Walter Ralph, 1932–
II. Series.
HF5548.2.A742   1984   001.53′5′024658   83-26648
ISBN 0-89391-220-4

Ablex Publishing Corporation
355 Chestnut Street
Norwood, New Jersey 07648

Contents

NATURAL LANGUAGE COMMUNICATIONS

AI MARKETS AND FURTHER APPLICATIONS

Preface

Each year, the Computer Applications and Information Systems area in the Graduate School of Business Administration at New York University organizes a symposium on new developments in computer applications of interest to business (topics for other years include decision support systems, human factors, and database methods and applications).

This book makes available papers presented at the 1983 NYU Symposium on Artificial Intelligence Applications for Business. The Symposium was held on May 18–20, 1983, at the NYU Graduate School of Business Administration in New York City.

The book is intended for researchers, practitioners, and members of the business and financial communities interested in artificial intelligence applications for business. The papers describe important existing applications, and the prospects for the next three to five years. The applications areas include expert or knowledge-based systems, natural language interfaces, robotics, and intelligent database software. The authors are well-recognized experts in their areas, and the papers are tailored specifically for those interested in practical applications.

The 1983 symposium itself was the work of many people. Once the CAIS faculty had decided upon artificial intelligence applications as the basic theme, the concept was developed by a committee consisting of Gadi Ariav, Jim Clifford, Jon Turner, Yannis Vassiliou, and myself. Charlsetta Langston and Carole Larson did the detailed planning, and for eight months they managed all of the innumerable arrangements and problems. Hank Lucas, the CAIS chairman, provided support and encouragement for all of us involved in the project.

When we discovered early in May that we were about to be over-whelmed by runaway success, Ariav took charge of the multimedia audio-visuals; Vassiliou and Margi Olson set up an impromptu print shop; Clifford played roving back; and Jon Turner, doing his thing, arranged for the participants some of the finest gourmet dining this side of the Atlantic. It may be true, as they say, that those who can't, teach. But at the Graduate School of Business Administration, some of those who teach certainly also can do.

The editing of the book was completed after I joined BBN, and I would like to thank Ray Nickerson, Senior Vice President at BBN Laboratories, and Dave Walden, President of BBN Laboratories, for making it possible for me to finish the work. I'm also indebted to Beverly Tobiason, who knew where things were and got them out.

Most of all, I want to thank the authors of the individual chapters. Not only have they provided an excellent set of papers, but rarer still, they provided them on time. And despite the shock this must have caused them, Karen Siletti and the Ablex people have produced the book efficiently, pain-lessly, and on schedule. For this we also must thank my good friend Mike Ginzberg, who invited us to publish with Ablex.

We hope you find the result a readable, informative, and useful book.

Walter Reitman
Cambridge
November, 1983

Artificial Intelligence Applications for Business: Getting Acquainted

1

Walter Reitman
ARTIFICIAL INTELLIGENCE DEPARTMENT
BBN LABORATORIES
CAMBRIDGE, MA 02238

In finding out how artificial intelligence can be applied in business, it helps to begin with some background. What is artificial intelligence? Who does it, and why? How did it get to where it is today? What kinds of problems does it deal with? What has it produced so far?

The business world has become interested in artificial intelligence both as a new way of approaching old problems, and as a tool for doing things that could not have been done before. The best known examples are natural language interfaces and expert systems. But other, less visible, applications may prove equally important. We give some representative examples, and then ask about the lessons to be learned from what has been achieved so far.

1. Introduction to Artificial Intelligence

This chapter outlines a few main themes, goals, and historical trends in artificial intelligence (AI), and makes some educated guesses about how AI applications for business are likely to develop over the near future. The chapter is meant to provide background for those making their first acquaintance with AI, and to introduce the chapters that follow (see Figure 1).

One thing to realize about AI is that it is not really so novel as today's newspaper and magazine coverage might suggest. The first published AI papers appeared almost thirty years ago, and interest in AI applications has waxed (and waned) several times since then. Current attempts to develop AI applications certainly are much broader and more intensive than any previous ones, and though reach may once again exceed grasp, no one this time expects AI to revert to its former status as an academic curiosity. Yet if AI applications are here to stay, it is not primarily because of any radical new achievements in AI itself. All the hard old problems are still

---

**Getting Acquainted With
Artificial Intelligence**

- Who does it, and why?
- How did it develop?
- What has it produced?
- Why is business interested in AI?
- Staffing and funding.
- What happens next?

Figure 1.

there. Instead, the conditions have changed. The universities now have produced sufficient numbers of skilled, experienced AI researchers to make business investment in human AI resources feasible. Senior people in the field have become deeply involved in the marketing and funding of AI applications. Radically decreased computing costs now make AI applications economically viable. And finally, as this symposium and others like it demonstrate, we are seeing what in the language of the seventies would be termed a general raising of consciousness in the MIS, business, and financial communities about the potential benefits of more intelligent hardware and software. For many very practical people, AI is no longer science fiction; like space, it has begun to be part of the real world.

The next thing to realize about artificial intelligence is that it is a child of very mixed parentage (see Figure 2).

Scratch an AI applications knowledge engineer and you may find a computer scientist with a background in the theory of algorithms, a psychologist trying to understand how people see patterns or solve problems, a linguist interested in subtle points of grammar, or even a logician-philosopher whose ambition is to express all knowledge in the first-order predicate calculus. Add to this the normal variations in human temperament, from the wild-eyed optimist who sees the distant future around each corner, to the pessimist who can't see anything going anywhere, ever, and you begin to see why the AI applications arena is as lively and confusing as it is.

Who Does Artificial Intelligence, And Why?

- Computer science
- Cognitive psychology
- Linguistics and philosophy
- Applications engineering

Figure 2.

Developmental Stages

- Stage 1: Conceptual Analysis and small systems, 1955–69
- Stage 2: Prototype systems, 1970–80
- Stage 3: Commercial applications, 1981—

Figure 3.

In trying to understand AI, it also helps to know a bit about how it has developed. Rick Hayes-Roth's chapter diagrams one view of the evolution of knowledge engineering in some detail, but for our purposes here the simple three-stage breakdown shown in Figure 3 will be sufficient.

The overwhelming fact of life of the first stage, from 1955 through about 1969, was the limitation on computer resources. In a day when 4000 words of core memory was all you had, it is easy to understand why conventional computer scientists shook their heads at the first list-processing language, which sacrificed as much as half the available memory in order to link up data elements into free-format dynamic structures. Because of computational restrictions, limited systems were all you could hope for. The trick was to decide whether any given system blazed the beginning of a potentially promising trail, or instead led to a dead-end.

Though these limited systems usually had little practical value, they did serve two important purposes. They demonstrated the feasibility of artificial intelligence, and they provided ground for some very sophisticated analyses of basic issues.

Stage 2 in the evolution of artificial intelligence saw the first significant prototype systems, notably in the area of natural language processing. Winograd's system was restricted to a simple toy world, and Woods's LUNAR program was limited to data on the composition of moon rocks. But within their restricted domains, both were remarkably good at understanding and answering quite complex questions in ordinary English. As these examples suggest, much of the technical base for today's stage 3 applications was laid down during the 1970s.

2. Results So Far

One approach to understanding artificial intelligence is to ask what it has produced so far. One answer, in very broad outline, is given in Figure 4.

We may begin with LISP. Adrian Walker's chapter argues very persuasively for PROLOG, a logic-based language, as an alternate vehicle for AI research. But at least for the present, most AI researchers would feel lost without LISP.

What Has Artificial Intelligence Produced So Far?

- Languages and programming environments
- General methods, concepts, and technical tools
- Explorations of fundamental problems
- Vehicles for representing complex, nonnumeric knowledge
- Working systems

Figure 4.

LISP has several features that make it a natural tool for AI work. It allows unconstrained tree structures. Thus the programmer never need specify, or even know in advance of execution, exactly what his or her data structures are going to look like. LISP was designed to facilitate the symbolic or nonnumeric data processing so essential to artificial intelligence systems. It uses the same basic representation for program and for data, thereby making it easy to treat one and the same data structure as something to be analyzed or to be run. LISP also provides a good basis for defining higher-order knowledge representation formalisms, e.g., frames. And, finally, as Beau Sheil's chapter demonstrates, the typical LISP installation provides a highly interactive and extremely powerful development environment. This makes it a great deal easier to explore and modify ideas and to develop new systems than is the case when working in a language like PASCAL.

Much less visible than LISP, but almost as important, are the general methods, concepts, and technical tools AI researchers have developed. These include procedural ideas, such as matching, backtracking, and generating and testing, as well as knowledge in depth about key technical problems and what to do about them.

Take the problem of dependencies, for example. If we are trying to improve our ability to play chess, we have to be able to relate the ultimate outcome of a game to events occurring much earlier in the course of play. If we are trying to program a planning system for a robot, it must be able to work its way back from dead-end plan outcomes to earlier hypothetical states of affairs. And if we are attempting to build an expert system that can explain its conclusions, it must be able to relate intervening states in the reasoning process to one another.

In the earlier stages of AI's development, such problems tended to be dealt with in isolation. Gradually, however, researchers became aware of the common elements underlying the particular cases. As they did so, they began to build up a body of concept, e.g., the notion of "truth maintenance," together with collections of generalized methods for dealing with the common problem in whatever context it appeared. This in-depth knowledge of general problems and what to do about them has become a basic element of the analysis and design aspects of "knowledge engineering."

Even in areas where much less progress has been made, AI investigators have developed a substantial informal understanding of the basic problems underlying the discipline. Learning and adaptation are a case in point. Although there are no generalized AI learning systems in any useful practical sense, investigators have learned a great deal about what the problems are, and have built up useful techniques for some interesting special cases. Here as elsewhere, basic and applied AI research are linked. The basic researcher needs to understand the hard problems in order to solve them eventually, and the applied researcher needs to understand them in order to avoid them while developing systems for immediate practical application.

One of the important ways in which AI systems differ from conventional computing is that they usually deal with "knowledge" in addition to "data" or "information." And one of the most significant outcomes of AI research is the collection of vehicles it has created for representing knowledge. These include logic-based formalisms, rule-based production systems, frames and schemas, procedural representations, and even analogical vehicles. What exactly these terms refer to in any particular case is a nice problem for a Ph.D. dissertation, and not something we want to get into here. Chandrasekaran's chapter, however, tackles a related practical problem, the question of how one goes about selecting from the repertoire of knowledge representations the most likely to be appropriate for a given applications problem.

3. AI *Applications Areas For Business*

For many, of course, the most interesting and important results to come out of AI are the individual applications possibilities it has produced. We now turn to these.

Howard Austin's chapter provides a detailed discussion of the extent of business interest in AI applications, and so all we will do here is mention some representative examples and indicate what makes them attractive (see Figure 5).

Probably the most visible AI applications programs are the expert systems. Austin, discussing his experience on the dip meter expert project at Schlumberger, points out that a really expert expert is by definition a rare bird, and from his employer's point of view he has drawbacks. He can get sick, or die, or be bought off by a competitor. Thus, replacing the human expert with an expert program, something that will work reliably round the clock, and can be replicated at will, is obviously a very attractive idea.

Why Is Business Interested In AI?

- Programming experts: the Schlumberger example
- Automating limited natural language interactions
- Automating manufacturing and assembly: DEC's programs
- Providing better database access and control
- Augmenting decision support: scenario generation
- Robotics: beyond the paint-and-weld robot

Figure 5.

As Harry Pople's chapter explains, however, figuring out how to do that is much less obvious.

We referred earlier to two stage 2 AI systems that were able to handle natural language inquiry. These days, the building of natural language systems and interfaces for commercial use has become another quite active AI applications area. These systems use a variety of approaches. Rusty Bobrow and Madeleine Bates describe what is available now, or likely to be available over the near term, and then provide some guidance for those interested in exploring the possibilities. Then Steve Shwartz takes a look at the same general area, but from the viewpoint of someone in the business of building natural language systems for commercial use. Finally, Bonnie Webber provides a broader, longer term view of natural language applications and suggests some ways in which the area is likely to develop.

Many expert systems have been built, but few have been exposed to the kind of intensive in-house testing given R1, DEC's VAX configurer. R1 is one of the very few expert systems that actually work day in and day out in an industrial setting, with no ifs, ands, or buts. John McDermott draws on his R1 experience as well as more recent work at DEC to provide some general suggestions about implementing expert systems, and Dennis O'Connor provides a complementary presentation, looking at DEC's expert systems in terms of the managerial issues involved.

Database formalisms sometimes are viewed as knowledge representatation schemes, but they are much more restricted, and restrictive, than the AI varieties mentioned above. Furthermore, if they are to be used really effectively, most of the intelligence typically has to be supplied by the user. But current database technology is a very well-established part of today's MIS context. Thus it becomes interesting and important to ask, as the chapter by Matthias Jarke and Yannis Vassiliou does, how AI techniques might be applied to improve database usage and performance.

Decision support systems are another prominent feature of the current MIS landscape. They range from simple Visicalc models all the way to the huge integrated systems used by large corporations to assist in capital budgeting and planning (Ginzberg, Reitman, & Stohr, 1982). Once again, however, such systems suffer from some severe restrictions, and the chapter by Dick Duda and Rene Reboh provides an opportunity to reconsider some of the issues of decision making in AI terms, in the light of their experience with the PROSPECTOR system.

Finally, no survey of the reasons for business interest in AI would be complete without some discussion of robotics. In particular, future developments in robotics are likely to draw very heavily upon AI research in perception and planning. Ruzena Bajcsy's chapter provides an overview of current robot capabilities, and then goes on to discuss what we may expect to find in the next generation of robots, and how these more powerful systems may be used.

4. Trends for the Future

Every discussion of AI applications should have a few guesses about the future, so here are mine (see Figure 6).

As for funding, we can distinguish three distinct sources: (a) Large corporations will increasingly fund AI groups, either as distinct entities, or else as part of expanded and redefined information systems staffs. Schlumberger and Atari are good early examples. (b) We also will see more AI software houses, backed by venture capital from various sources. These will be staffed with smart systems people who have the advantage of their familiarity with the new AI concepts and tools. Remember, however, that in the last analysis they are building and marketing software. Thus we should expect them to follow the same developmental paths and be subject to the same market forces and market opportunities as other software developers and suppliers. (c) The last main source of funds is the government, in particular through such agencies as DARPA and the R&D funding units of the military. The amounts and the problems targeted no doubt will vary with changing political and economic conditions, and the relative magnitude of government funding may decrease as the private sector funding of AI ap-

---

What Happens Next?

- Funding
- Understanding the strengths and weaknesses: expert systems
- Toward intelligent assistants
- The reemergence of no-holds-barred AI systems
- Unattributed diffusion: Xerox Star, Lisa, and "softer software"

Figure 6.

plications grows (see Howard Austin's chapter). But in view of the perceived importance of AI to the military, and to the national economy generally, we should expect this source to stay strong.

As experience with various types of AI applications grows, so should our understanding of their strengths and weaknesses. When we consider a potential expert system applications, for example, we should now know enough to ask at least the following questions. (a) How rare are solutions to the problem, and how are they distributed? (b) How decomposable is the problem? (c) How formal(izable) is the knowledge? How much is rule-based? (d) How difficult is it to acquire the knowledge? (e) How much noise, or error, is there in the knowledge base? R1, for example, may work as well as it does because acceptable VAX configurations are not particularly rare, because the configuration problem is highly decomposable (at least once you have figured out how to do it), and because all of the requisite knowledge can be represented in rules. The development of sophisticated knowledge about the strengths and weaknesses of a particular AI applications technology is likely to prove as important as the development of the technologies themselves, and the chapters by Chandrasekaran, Duda and Reboh, McDermott, and Pople will be particularly useful in this regard.

Recognize that the AI applications areas of today are surely only the tip of the iceberg. At research centers around the world, AI investigators are rethinking the whole question of management information systems, how they should be developed, and what they should be expected to do. The chapters by Roger Schank, Beau Sheil, and Bill Woods illustrate some of the kinds of work now under way. Progress toward intelligent computer assistants will be slow, because we do not yet have general solutions for alll of the basic problems. But no one should doubt that it will come.

When they do come, these more powerful AI-based systems are likely to be much more complex and varied in form. Rule-based systems captured the public imagination because they were relatively easy to understand and implement. But many kinds of knowledge and process are difficult to shoehorn into a simple rule-based framework, and as AI achieves respectability in the business community, less need will be felt to market it in simple packages. The final chapter of this book gives some practical suggestions to those business people responsible for acquiring and managing an AI applications capability.

Finally, as AI ceases to be a controversial laboratory curiosity, the link between AI applications and conventional software will blur. We can see this happening already, witness the substantial AI influence on the software for the Xerox Star and the Apple Lisa. We quote from a recent *New York Times* column on technology to show how this unattributed diffusion occurs.

> If metaphor, integration and networking are this year's buzzwords, next year's might be "soft software," according to William H. Gates, the

chairman of Microsoft, a leading producer. Just as software takes a general purpose computer and adapts it to a particular task, so soft software will take a standard product and adapt it to a particular user.

The "softer" computer programs will be written to understand what the user wants, rather than what he says, and to forgive such things as typing errors in entering commands. Like a human assistant, a computerized filing system might become progressively faster at retrieving frequently requested data, rather than treating each new request as a brand new task.

In other words, softer software will make for more personal personal computers.

What is interesting is that there is not a single reference to AI in the entire piece, and yet the specifications Gates describes clearly are AI-inspired. Researchers who have spent years struggling to get AI taken seriously may have a hard time adjusting to suddenly having it taken for granted. But they could hardly ask for a clearer sign that AI applications are here to stay.

5. References

M. J. Ginzberg, W. Reitman, & E. A. Stohr (Eds.). Decision support systems. Amsterdam: North-Holland, 1982.

New York Times, Feb. 17, 1983, p. D2.

Building Expert Systems*

John McDermott
DEPARTMENT OF COMPUTER SCIENCE
CARNEGIE-MELLON UNIVERSITY
PITTSBURGH, PA 15213

Though only a handful of expert systems have as yet found a home in industry, patterns in their developmental histories are beginning to emerge. This chapter describes four stages that developers of expert systems typically pass through and then exemplifies these stages with three case studies.

1. Four Stages in the Life of a Knowledge Engineer

The process of building an expert system ordinarily consists of the following four stages:

- Task definition
- Initial design
- Earnest knowledge extraction
- The aftermath, technology transfer

If a knowledge-intensive problem has been identified, the first stage in building a system to assist in solving the problem is to delimit its task; since solving knowledge-intensive problems is often a communal endeavor and since the problems are ordinarily not cleanly bounded, there is frequently a real issue of determining just what role the expert system is to play. Once the task has been defined, the developer must create an initial design for the system on the basis of the expected characteristics of the as yet uncollected domain knowledge; the trick here is to figure out how

* Many people have participated in the design and development of R1, XSEL, and PTRANS. Those of us at CMU who have worked on these systems appreciate the help we have received from people at Digital in Lou Gaviglia's and Dennis O'Connor's organizations.

that knowledge can best be exploited to guide the system's problem-solving behavior. Once there is a design, the knowledge extraction process will have a focus; thus the skeletal system implicit in the design can begin to be fleshed out with knowledge. At some point, long before the system has become expert in its task, it will be able to provide useful assistance; at that point issues arise of how to continue to develop the system in an operational environment.

In order to appropriately delimit a task, it is necessary to define the role the expert system will play relative to others who perform complementary tasks. If there is a large amount of knowledge that is relevant to some problem, then it is often the case that the required expertise is distributed among several people. Each individual's role is defined both by his or her area of competence and then by a level of competence within that area. When an expert system is developed to assist with some problem, the first question that has to be answered is whether the system will assume an existing role or carve out some new role. How that question is answered will have strong impact both on how knowledge acquisition is handled and on how transfer to an operation environment is carried out.

The stage in building an expert system that is least well understood is that of creating an initial design. What sets an expert system apart from other computer programs is that it is able to perform its task because it can bring large amounts of domain knowledge to bear at each step; this knowledge gives the expert system its problem-solving power. But in order for an expert system to be able to exploit its knowledge, the knowledge must be represented in such a way that all of it that is relevant at any step is immediately available to the system. The principal problem in developing an expert system is in determining how to represent the knowledge so that it is always available when needed. In order to extract knowledge from experts, it is first necessary to understand how that knowledge is going to be represented; but without already having that knowledge, it is not clear what has to be represented. At present, all we can do in this circumstance is fall back on general arguments for representational adequacy. In the future, we will hopefully have enough of an understanding of the representational demands of a wide variety of task types that the particular representational demands of some new task will be clear on the basis of the similarities of that task to one or more of those task types.

All of the expert systems that have been developed to date have acquired their knowledge over a long period of time. In the typical case, a relatively small but central fraction of the domain knowledge has been collected by a knowledge engineer interacting over a period of a few weeks or months with a domain expert. This knowledge has been used to build a novice system that can behave appropriately in commonly occurring situations, but that is almost totally insensitive to special circumstances. Knowledge is then extracted from an expert by asking the expert to observe the behavior of the novice, point out mistakes the novice makes, and for every mistake, indicate what knowledge the novice is lacking or has wrong. Over a period of months or years, this process uncovers enough of the domain knowledge so that the novice grows to be an expert. Variations in this process occur depending on the source of the expertise. If a system is to assume an existing role (i.e., if there are human experts who perform the same task), then the source of the knowledge is those who play that role. If the system is to carve out a new role, part of the knowledge acquisition problem is identifying the sources of knowledge.

A human expert does not stop acquiring knowledge of his domain at some arbitrary point in his life; there is always room for him to become more expert. Likewise, the development of an expert system is never finished. Thus expert systems have no need for maintainers, but they do always need developers even after they have moved to an operational environment. This need poses a problem since the current demand for people with some background in artificial intelligence far exceeds the supply. If expert systems are to have an appreciable impact on industry, ways must be found to provide a significant number of people with an understanding of AI techniques.

Since 1978, Carnegie-Mellon University and Digital Equipment Corporation have been collaborating on the development of a number of expert systems. Three of these systems are now being used at Digital. R1, a computer system configurer, has been used on a regular basis since 1980. XSEL, a salesperson's assistant, and PTRANS, a manufacturing management assistant, have just recently begun to be evaluated. XSEL (McDermott, 1982a) helps the salesperson select a set of components tailored to a customer's needs and design a floor layout for the resulting system. Given the set of components selected with the help of XSEL, R1 (McDermott, 1982b) adds any missing support components and determines what the spatial relationship among the components should be. The XSEL user can call on R1 to configure or reconfigure the current set of component selections and can provide information to XSEL which insures that the resulting configuration will satisfy special requirements of individual customers (McDermott, 1981). PTRANS (Haley, et al., 1983) helps insure that the inevitable problems which arise and threaten to delay the delivery of systems to customers are resolved satisfactorily. In time, XSEL will interact with PTRANS so that a delivery date can be confirmed as soon as an order is booked.

All three of these systems are implemented in OPS5 (Forgy, 1981). OPS5 is a general-purpose rule-based language; like other rule-based languages, OPS5 provides a rule memory, a global working memory, and an interpreter which tests the rules to determine which ones are satisfied by the descriptions in working memory. An OPS5 rule is an IF-THEN statement consisting of a set of patterns which can be matched by the descriptions in working memory and a set of actions which modify working memory when the rule is applied. On each cycle, the interpreter selects a satisfied rule and applies it. Since applying a rule results in changes to working memory, different subsets of rules are satisfied on successive cycles. OPS5 does not impose any organization on rule memory; all rules are evaluated on every cycle.

There are informative similarities and differences in the developmental histories of R1, XSEL, and PTRANS. Each of the following sections is a case study of the development of one of these systems. Since all three of the systems were developed at the same research institution and since they are all being used by the same company, some of the similarities and differences are undoubtedly due to local peculiarities. For discussions of other expert systems which may help to broaden the picture presented here, see Feigenbaum (1977) and Stefik et al.,(1982).

2. *Building* R1

Digital differs from most computer manufacturers in the degree of flexibility it allows its customers in component selection; rather than marketing a number of standard systems, each with a limited number of options, Digital markets processors with relatively large numbers of options. One of the results of this marketing strategy is that many of the systems it sells are one of a kind, and consequently each poses a distinct configuration problem. A computer system configurer has two responsibilities: he must insure that the system is complete (i.e., that all of the components required for a functional system are present), and he must determine what the spatial relationships among the components should be. Because a typical system has 100 or so components which have a large number of different possible interrelationships, a significant amount of knowledge is required in order to perform the configuration task competently.

Before the advent of R1, the configuration task was performed by people of varying degrees of expertise at various points in the order processing pipeline. Each order was examined by people known as technical editors at least twice. The first examination was to insure that the order was complete (i.e., that the particular set of components on the order were sufficient for a functional system). The second examination, though concerned with completeness, also had the purpose of specifying what the relationships among the components should be. Because the technical editor's task is fairly tedious, it is not a position that engineers aspire to; thus the people who have this job tend to have fairly weak engineering backgrounds. Technical editors can, however, call upon an engineer who understands computer system issues whenever they encounter a problem that is beyond their competence.

2.1. *Task definition*

Several factors influenced the initial definition of R1's task. First, it was simply assumed that R1 would perform the task that the technical editor responsible for specifying the relationship among components performs. A system which simply checked for completeness would have been of use, but at that time, Digital's primary concern was with finding a way to insure consistency among configurations. Second, the possibility of developing a system that could reason like an engineer when encountering a previously unencountered problem was never considered; as a result, R1's knowledge turned out to be the surface knowledge of the technical editor, rather than the deeper knowledge of the engineer. Third, since a technical editor is ordinarily responsible for a single system type (e.g., a PDP-11/44, a VAX-11/780), it was assumed that R1's initial task would be to configure just a single system type. But it was not at all clear which system type was the right one to start with. The alternatives were either one of the PDP-11 systems or the VAX-11/780. The main reason for favoring a PDP-11 was that configuration was more of a problem for the PDP-11 systems. The main reason for favoring the VAX-11/780 was that it was easier to configure, primarily because the number of supported components was, at that time, so much smaller. It was decided that the initial task should be VAX-11/780 configuration. As it turned out, this was a fortunate choice since it allowed R1 to become a good novice configurer more quickly and as a result gain increasing amounts of cooperation from those people who had knowledge it needed.

2.2. *Initial Design*

After we had some understanding of what was involved in the configuration task at an abstract level, but before we had more than a few examples of the knowledge technical editors have in their heads, we made two basic design commitments which implicitly defined the structure that R1 would have. One of these was to encode the domain knowledge as situation/action rules; the other was to view the configuration task as a collection of loosely coupled subtasks. Each of R1's rules defines a situation in which some particular configuration step is required. Since a rule is applied only when the descriptions in working memory satisfy the condition part of that rule,

R1's rules do not specify sequences of steps to be performed, but rather specify the possible steps without committing in advance either to particular steps or to the sequence in which the steps should be performed. R1 relies on this characteristic of rule-based systems in two ways: (a) because there is so much variability in the steps required for configuring different systems, enumerating all possible sequences of steps would be an almost impossible task; encoding knowledge in rules permits each small piece of situation-specific knowledge to recognize when it is relevant and thus dynamically determine the appropriate next step, and (b) because the knowledge that R1 needs in order to become an expert configurer can only be acquired over time, new knowledge must be continuously assimilated. Since a rule defines the type of situation in which it is relevant, rules do not interact very strongly and thus adding new rules is straightforward.

To say that the configuration task can be viewed as a collection of loosely coupled subtasks means that each piece of configuration knowledge can be associated with a particular subtask. When the nature of a task allows such partitioning of knowledge, issues of control are ordinarily easier to resolve. Because rule-based systems are data-driven, it can happen that rules which are satisfied but have no relevance to the subtask currently being performed are applied before (or even instead of) rules that bear on the current subtask. We decided to avoid this problem in R1 by including, in the condition part of each rule, a description of the particular subtask that the rule is relevant to. When a rule (relevant to a particular task) recognizes that some subtask needs to be performed before the current task can be completed, that rule asserts a description of the subtask which needs to be performed.

2.3. Knowledge Extraction

After one week's interaction with a Digital engineer, it took about three months to develop an initial, primitive version of R1. This novice version had about 250 situation/action rules; this amount of knowledge enabled R1 to configure very simple VAX–11/780 systems without making any mistakes. Over a period of about five months, expert configurers observed R1's performance and indicated what knowledge R1 was lacking; this knowledge was encoded as rules and added to R1. A total of about one man-year was spent in getting R1 to the point where Digital could formally evaluate it. At 750 rules, R1's performance was fairly impressive; the experts concluded that it had almost all of the knowledge it needed, so it began to be used by Digital to configure all VAX–11/780 systems. R1 now has almost 2500 rules. Some of this additional knowledge was required to enable R1 to configure systems other than the VAX–11/780. But much of it is knowledge that was simply overlooked by the experts. The experts did not think to supply this knowledge until R1 tried to configure systems which could not be configured correctly without it. Because some of this knowledge is very rarely needed, it has taken three years to uncover.

2.4. Technology Transfer

As R1 was being developed at CMU, little attention was paid to the question of who at Digital was going to maintain and continue to develop R1 once it moved to Digital. R1 moved to Digital in January of 1980, and a small group of programmers was given the responsibility of seeing that R1's potential was realized. For about a year, the people at CMU who had developed R1 provided a significant amount of assistance. But by the end of 1980, the group at Digital had became sufficiently familiar with R1 and with the AI techniques it embodies to be able to proceed without additional assistance from CMU. During the past three years, the people at Digital have extended R1 from a 1000 rule system to a 2500 rule system.

The number of people who have worked directly on R1 has, over the past two years, remained at about ten. Three of these people are engineers who are responsible for identifying the configuration knowledge R1 lacks on the basis of mistakes it makes and for adding new product information to the database. Four of the people are programmers; the engineers tell them what configuration knowledge R1 is lacking and they determine how to represent this knowledge in the form of rules. The other three people are R1's interface to the user community. Considering that none of these people had any background in AI when they began their jobs, they have accomplished a great deal.

3. Building XSEL

The degree of flexibility Digital allows its customers in component selection also has the consequence of placing a significant burden of design responsibility on salespeople. In order to select a set of components that will satisfy a customer's needs, the salesperson must isolate all of the constraints that the customer's needs impose and then find a set of components which satisfy those constraints. The task of finding an appropriate set of components given some constraints is fairly straightforward. More difficult is the task of defining the needs in a fashion that allows judgments of the relative importance of various constraints to be made.

3.1. Task Definition

R1's task was defined by determining what the responsibilities of the technical editor were and then defining R1's task in terms of those responsibilities. The case with XSEL is quite different. The problem which XSEL addresses, the sizing problem, is partly the responsibility of salespeople and partly the responsibility of special sizing consultants. XSEL has some of the responsibilities of both. Given a statement of the intended use of a computer system, XSEL's task is to determine precisely what components that system should consist of. To perform this task, it must have the following capabilities: (a) It must understand how to convert indirect measures of computing resource needs (e.g., number of customers, billing cycle) into direct measures (e.g., megabytes of memory, lines per minute of output). (b) It must understand how to convert direct measures of computing resource needs into appropriate quantities of specific components. (c) Finally, it must understand the constraints that the room or rooms that will house the system impose on the selection of components. Since the first of these capabilities is the most difficult to realize in a program, and since the other two, by themselves, are of value to the salesperson, it was decided that the initial version of XSEL would provide only those two capabilities. This would allow the user community to become accustomed to the tool while the indirect to direct capability was being developed.

3.2. Initial Design

There are four design commitments which implicitly define XSEL's structure. Two of the commitments are those we also made for R1. The third was to insure that an any point in an interaction, any piece of information previously entered or any conclusion made by XSEL could be modified by the user; the fourth was to require that XSEL be able to explain all of its sizing decisions. A primary function of XSEL is to provide the user with a means of exploring the implications of various characterizations of a customer's needs. Since it is impossible to determine precisely what a customer's needs are likely to be and since the relationship between computing resources and needs is never very clear cut, it is important that the user be able to easily modify both the descriptions of needs and the implications drawn from those descriptions. For much the same reason, it is important that the user be able to get satisfying explanations of how various conclusions were arrived at. Since various users will have different degrees of understanding of sizing issues and since the strengths of XSEL's conclusions depend on a variety of often somewhat obscure factors, it is important that the explanations that XSEL gives are able to be tailored to the situation at hand.

3.3. Knowledge Extraction

The most interesting difference between the development of R1 and the development of XSEL is in the way in which the knowledge that the systems have was acquired. With R1, experts identified the mistakes that R1 made because of insufficient knowledge and supplied that knowledge. With XSEL, the case is somewhat different because for much of what XSEL does, the issue is not whether it is correct or incorrect, but rather whether it is helpful or unhelpful. This difference arises because R1 does not assist anyone in performing a task, but simply performs the task. XSEL on the other hand, is an assistant. In order to provide an environment in which the developers of XSEL could find out precisely what knowledge XSEL needed, Digital, early in XSEL's development, formed a user design group consisting of about 20 salespeople from across the United States. The members of this group have been XSEL users since the group was formed. If, when using XSEL, they have comments either about its helpfulness or its correctness, they make those comments via XSEL to the developers; they also meet every three months for two days to discuss the system's progress. When the initial version of XSEL was handed over to Digital a year ago, it had about 2000 rules; this represented approximately 4 years of effort. Since that time, Digital has added 500 rules to that initial version and we at CMU have developed a set of 500 rules that begin to give XSEL the ability to convert indirect measures of computing resource needs to direct measures.

3.4. *Technology Transfer*

Digital was somewhat better prepared to accept XSEL from CMU than it was to accept R1. Since the R1 development group was already in place, it was the obvious group to take over the development of the initial version of XSEL. The transition has not, however, been as smooth as one might have hoped. Because R1 now plays an integral role in Digital's manufacturing organization and because there is considerable pressure on the R1 support group to extend R1's capabilities to additional system types, the people responsible for R1 have more than enough to do. Because XSEL is still only being tested, its claim on those knowledgable resources is very weak; consequently, the XSEL support group that was formed consisted, like R1's group, of five people, but only one had had any experience with rule-based systems. Thus there has been a learning period for the XSEL developers, just as there was for the R1 developers, and so XSEL has progressed less rapidly than it otherwise might have.

4. *Building* PTRANS

Digital's final assembly and test plants build complex systems from components produced by high volume plants. The principal management tasks are determining when and where on the floor to build each system, insuring that the necessary parts are on hand when it is time to issue them to the floor, and tracking the progress of each system on the floor so that problems can be resolved as they arise. In order to perform these tasks, managers must be able to construct plans and then modify the plans appropriately as unforeseen circumstances arise.

4.1. Task Definition

Like XSEL, PTRANS is an assistant, and thus defining its task required finding out what forms of assistance would be helpful. Unlike the sales task, however, the manufacturing process management task is a collection of quite diverse subtasks, each of which is typically performed by a different person or a different group of people. This suggested that PTRANS should be a group of cooperating assistants. Each assistant would, like XSEL, be responsible for making someone more capable, but would also be responsible for knowing what other assistants it should communicate with under what circumstances.

PTRANS differs from conventional manufacturing support systems in that its primary focus is on helping to manage exceptions. In conventional systems, the primary focus is on helping to characterize resource requirements so that adequate plannning can keep the number of problems that arise to a minimum; much less attention is paid to providing assistance with the problems that inevitably do arise. PTRANS, too, tries to provide enough visibility to potential problems so that they can be avoided. Most of its knowledge, however, bears on ways of resolving problems once they have arisen.

4.2. Initial Design

Three of the four design commitments which implicitly define XSEL's structure also define PTRANS' structure. PTRANS' knowledge is encoded as situation/action rules, any of its data or conclusions can be modified by the user, and it can explain all of its decisions. The fourth commitment, however, that the task be viewed as a collection of loosely coupled subtasks, turned out to be inappropriate. PTRANS must be able to perform two different sorts of tasks simultaneously: (a) it must construct and then guide the implementation of build-plans (i.e., must assist in determining when, where, and how each system should be built), and (b) it must recognize events which make implementing a build-plan infeasible and make appropriate modifications to that plan. Events that signal problems may, either directly or indirectly, affect decisions that need to be made about systems that are currently being worked on. Therefore, if the problems are not noted as soon as they arise, any recommendations PTRANS makes will be valid only in a world that no longer exists. In order to recognize problems when they arise, PTRANS must have immediate access to a great deal of its knowledge. This problem is addressed in PTRANS by providing it with knowledge at two levels. Part of its knowledge is associated with subtasks in the same way that R1's knowledge is. But above that level of task-specific knowledge, there is a level comprised of a large number of demons. Because OPS5's recency strategy gives preference to rules that are satisfied by the most recent working memory elements, any satisfied demon rules will always be applied before rules associated with a subtask. Organizing knowledge in this way, gives PTRANS almost immediate access to all of its knowledge.

4.3. Knowledge Extraction

Approximately 4 man-years of effort have gone into the development of the initial version of PTRANS; it will be handed over to Digital shortly. PTRANS now has about 1800 rules; about 900 of these rules are demons. Of the approximately 900 demon rules, about 400 recognize that required information is missing and supply it. About 425 recognize (possible) inconsistencies and either resolve them or propagate their consequences. About 75 recognize inconsistencies which are complex enough to require a subtask to be generated. In order to determine what knowledge each of PTRANS' cooperating assistants would require, a group of seven experts, one for each of the subtasks that PTRANS knows how to assist with, was formed; that group has to this point provided all of the knowledge that PTRANS has. When PTRANS moves to Digital, this group will be responsible for identifying and supplying missing knowledge.

4.4. Technology Transfer

When Digital took over responsibility for further developing XSEL, there were people at Digital who could have done the job ably. Unfortunately, all of those people were needed for the further development of R1. To avoid that problem with PTRANS, Digital started training people before they were needed. Over the next three years, several Digital employees will each spend from nine to twelve months at CMU learning about AI techniques by assisting in developing expert systems. Others will spend a comparable amount of time working in the groups at Digital that are responsible for the expert systems already in use there. Digital's expectation is that at the end of that three year period, they will have a resource that will enable them to support all of the expert systems in use within Digital as well as to design and develop new systems.

5. Concluding Remarks

There is now a set of AI tools that make it relatively straightforward for the initiated to develop expert systems; though the tools will surely improve substantially over the next few years, useful systems can be built now. But this does not mean that expert systems are going to become widespread soon. There is a serious technology transfer problem. Because expert systems are continually in need of further development, in order for them to become commonly used tools tens of thousands of people who can assist with their development will have to be found. At the present time there are probably no more than a few hundred people in the world who have the necessary experience. We are years away from the point where expert systems will be able to be developed by the uninitiated. And currently the few places that provide the required training do not have the resources to train more than a few people at a time. Thus it will be a while before expert systems are everywhere.

6. References

Feigenbaum, E. A. The art of artificial intelligence: themes and case studies in knowledge engineering. In *Proceedings of the Fifth International Joint Conference on Artificial Intelligence,* 1977, 1014–1029.

Forgy, C. L. *OPS–5 user's manual* (Tech. Rep. CMU-CS-81-135). Department of Computer Science, Carnegie-Mellon University, 1981.

Haley, P., Kowalski, J., McDermott, J., & McWhorter, R. *PTRANS: A rule-based management assistant* (Tech. Rep.). Department of Computer Science, Carnegie-Mellon University, in preparation, 1983.

McDermott, J., & Steele, B. Extending a knowledge-based system to deal with ad hoc constraints. In *Proceedings of the Seventh International Joint Conference on Artificial Intelligence,* 1981, 824–828.

McDermott, J. XSEL: A computer salesperson's assistant. In J. E. Hayes, D. Michie, & Y-H Pao (Eds.), *Machine Intelligence 10.* New York: John Wiley & Sons, 1982, 325–337. (a)

McDermott, J. R1: A rule-based configurer of computer systems. *Artificial Intelligence,* 1982, *19*(1), 39–88. (b)

Stefik, M., Aikins, J., Balzer, R., Benoit, J., Birnbaum, L., Hayes-Roth, F., & Sacerdoti, E. The organization of expert systems, a tutorial. *Artificial Intelligence,* 1982, *18*(2), 135–173.

Knowledge-Based Expert Systems: The Buy or Build Decision

Harry E. Pople, Jr.
DECISION SYSTEMS LABORATORY
&
GRADUATE SCHOOL OF BUSINESS
UNIVERSITY OF PITTSBURGH
PITTSBURGH, PA 16261

There appear to be quite a few organizations — in all branches of government, the military, business, industry and the service sector — that have begun to investigate the potential applicability of AI technology in their operations. As an indication of the growing interest in this field, a meeting early in 1983 at MIT on "the Reality of Artificial Intelligence" attracted around 900 participants from 150–200 organizations. This attendance figure surpassed that of any previous symposium sponsored by the MIT industrial liaison program, including the one a few years ago on biotechnology and genetic engineering.

Of course, the articles in *Business Week, Fortune,* the *Wall Street Journal* and other business, scientific and popular media have helped stir this degree of interest. Somewhat fortuitously, this increased awareness of the new technology has come at a time when many companies are desperate to find ways to increase productivity and in other ways gain a competitive edge. Unfortunately, the sometimes hyperbolic outpourings in the media have fed this hunger with the promise of incredible benefits to be gained through use of "knowledge engineering" and "expert systems" techniques.

As one of the few battle-scarred veterans in this business, I am often called upon to advise newcomers to the field. Based on the experience of many such encounters, I have become somewhat concerned that potential adopters of this new technology may be expecting too much, too soon. Because it may appear to the uninitiated that use of the "expert systems" approach has led to the development of computer programs that perform as well as competent professionals in science, medicine, law and other disciplines, there is the expectation that if only one can get a handle on this new technology, it ought to be possible to solve any number of what now appear to be intractable problems.

1. The Reality of Intelligent Systems

What *is* the reality of AI, and where does it enter into the systems design equation? There really are two answers to this. On the one hand, researchers in the AI field have developed a number of conceptual tools that make it relatively easy to structure and manipulate symbolic databases. The notion of representing information in the form of networks of nodes and links is one of the major contributions of AI research to computing generally. Additionally, techniques for searching such network data structures — particularly those implicitly defined networks that have the potential to grow in size exponentially, and without bound — have been a major concern of the AI tool builders. Another important idea that has emerged from AI investigations is the utility of a type of data structure called a "production rule" that can be used both to encode knowledge, and to define computations that can be expressed with a considerable degree of modularity.

These are examples of the types of tools that AI programmers tend to have at their fingertips, which are not generally well known in the systems analysis and data processing shops (although somewhat similar tools are to be found there: e.g., the decision table, long a staple in the data processing menu, shares many of the characteristics and virtues of the production rule). One of the benefits of introducing these techniques in any systems analysis and programming operation is that they tend to simplify the programmer's task, make for more modular organization of programs and data, and permit thereby the attack on more complex problems.

In the minds of many observers, however, use of these tools is seen as the defining criterion in determining whether or not a particular system is to be considered an "expert system"; indeed, the standard questions asked in assessing a new system often take the form "What inference engine does it use?" and "How big is its rule base?"

Is every system that employs these techniques to be considered an AI program? And what of systems that do not employ these techniques in any recognizable form; are they automatically to be relegated to the 'unintelligent' category? If we look only at the techniques used in organizing and designing the programs, we will be neglecting what is perhaps the most important aspect of artificial intelligence research, namely: use of methods of the cognitive psychologist to study how people identify, structure and solve tough problems.

From its earliest days, the field of artificial intelligence has adopted a research paradigm that involves the study of aspects of intelligent human behavior in the context of some significant problem domain (e.g., chess, symbolic logic, theorem proving, organic synthesis, medical diagnosis). Theories are developed to account for the observed behavior, drawing in part on the relevant models of the underlying scientific disciplines. In addition, to varying degrees, AI investigators make use of introspection — if they happen to have some expertise in the subject domain as well as in the fundamentals of AI — or more commonly, the introspective, retrospective, or "thinking aloud" reports of expert practitioners engaged in typical planning, problem-solving, or decision-making activities in the subject domain.

One of the side benefits of embodying these domain-specific theories of expertise in working computer models is the potential for application of the resulting programs as "expert systems." In many cases, empirical investigation has shown that the aspects of intelligent behavior captured by domain-specific AI models can indeed be helpful as decision or problem solving aids to practitioners in the domain. It would be a mistake, however, to view these domain investigations as "applied artificial intelligence"; instead, this constitutes the experimental component of AI research, having the potential to reveal important new theoretical insights bearing on one or more of the underlying disciplines of AI.

2. Artificial Intelligence as Experimental Computer Science

On the basis of my experience in this mode of AI research, I am inclined to characterize the modeling of expertise by a naive inquirer interacting directly with an expert as essentially an empirical process. One begins by acquiring a small number of facts, on the basis of which some hints may emerge as to the expert's structuring of knowledge. These structures in turn may provide cues as to the process by which this knowledge is accessed and used in the course of reasoning and problem solving. Further process cues may be obtained from direct testimony of the expert, from a study of protocols and the expert reasoning aloud, and from the inquirer's own introspections concerning the structure and process of knowledge in other domains.

The next step is to fashion from these empirical findings a testable hypothesis; and here, for the computer scientist, there is a considerable armamentarium including but not limited to the models and methods of AI that can be brought into play in the construction of a working model. It should perhaps be emphasized that what results from this stage is at best a model of the inquirer's concept, which may or may not bear much resemblance to that of the expert. Nonetheless, if reasonably faithful to the inquirer's emerging concept of the domain, the model can serve to guide and sharpen the further search for more subtle aspects of the expert's information structure and process.

There are a number of implications that flow from this characterization of the knowledge modeling process that I would like to summarize. First, the term "applications" often used to characterize this sort of investigation is something of a misnomer. It suggests the existence of a general theory which is being instantiated by a "knowledge engineer," whereas experience suggests that the modeling of expertise is primarily an experimental, theory-formation activity.

A corollary of this proposition is that the models and methods of AI, which often prove useful at various stages of the modeling process, should be regarded merely as tools of the investigator — not theories. One should not set about to fit a model of expertise to the models of AI. It would be better to devise new methods and techniques even at the risk of being called ad hoc if this is necessary in order to deal with the essential nature of the problem domain being investigated.

It also follows that one must be prepared to throw over any or all of a given model when further investigation reveals subtleties of expertise that cannot be represented within that framework. One must take whatever comfort there is in knowing that the model has served its purpose if it has sharpened the investigator's awareness of the structures and processes that underlie expertise, and then move on to the more sophisticated models and experiments enabled by this new level of understanding.

3. Buy or Build? First Identify the Real Problem!

One of the generalizations that might be drawn from the first decade of research into expert systems is that a competent performance program will take five to ten years to create, costing from one to two million dollars or more. Certainly if one looks at the developmental history of programs such as DENDRAL (Buchanan & Feigenbaum, 1978), MYCIN (Shortliffe, 1976), PROSPECTOR (Duda, et al., 1978), INTERNIST/CADUCEUS (Miller, et al., 1982; Pople, 1982) — these figures seem to be about par for the course. In light of this experience, it is understandable that newcomers to this field of endeavor may seek to bypass the heavy startup costs, and piggyback on this earlier investment by making use of the "inference engines" underlying these systems, abstracted from their original domain dependence and available off-the-shelf for adaptation to new applications.

While there are undeniably special cases where there are reasonable mappings between the structure of some important problem in a new domain and that underlying an off-the-shelf expert system, more commonly there are at most only suggestive parallels. In such cases, there may be a strong temptation to bend the problem to fit the chosen tool, or even to contrive a fictitious problem "for demonstration purposes." While this practice may be appropriate as a vehicle for training staff in the technology of knowledge based expert systems, it should not be mistaken for the real thing. To find acceptance by practitioners in the application domain, an expert system must be structured around one or more problems of real concern to those intended users.

The importance of identifying the right decision problem is illustrated by our experience in the design of decision systems for medical diagnosis, discussed in the following section.

4. Identifying and Structuring the Decision Problem

In the domain of medical decision support systems, there have been a number of approaches to computer aided diagnosis that have been developed over the past two decades. The primary features distinguishing these diverse approaches include:

1. the nature of the decision problem being dealt with,
2. the procedure employed as the basic problem solver, and
3. the allocation of responsibility to the user and to the computer for various aspects of the overall process.

The great majority of diagnostic programs assume a given decision problem. The premise underlying use of such procedures is that the clinical problem can be encompassed by a fixed differential diagnosis; i.e., there is a given set of diagnostic possibilities, and it is a presupposition in using the program that the patient has one and only one of the listed disorders. Another way of saying this is that the list of diagnostic possibilities is exhaustive and mutually exclusive, a necessary condition for the application of many statistical decision making strategies, notably Bayes's rule.

Such an assumption is unacceptable in the case of complex clinical problems having the potential for multiple diagnoses. There are, however, alternative approaches to the design of diagnostic programs in which the computer plays a more active role in the formulation of diagnostic tasks. The two principal methods that have been employed for computer-assisted evocation of diagnostic tasks are discussed in the following sections.

5. The Binary Choice Task Formulation Strategy

One way to arrange for the evocation of diagnostic tasks would be to use what in the terminology of artificial intelligence is referred to as the method of "hypothesis and test." Using this approach, the decision maker would hypothesize in some systematic fashion all possible diagnoses, and for each, create the task of determining whether this hypothesized diagnosis can be verified as contributing to the patient's illness. Such tasks would not be in the form of differential diagnoses (is the illness X? or Y? or Z?). Instead, each such task would entail a choice between the presence or absence of a particular disease (is the case X? or not X?).

This approach to task structuring has the virtue of ensuring that each task considered consists of an exhaustive and mutually exclusive set of alternatives (the disease either is or is not present, and it cannot be both). For this reason, it can be used in the design of Bayesian diagnostic procedures that are suitable for use in clinical problems having the potential for multiple concurrent diagnoses. (The hypothesis generation mechanism of MYCIN, for example, can be interpreted as a binary choice task formation strategy.)

Ordinary Bayesian task formulations, structured as differential diagnoses, require the assumption that one and only one of the diseases in the differential list is present in the case. This conventional problem structure entails one task and one decision, whereas a formulation based on a binary choice (between X and not X) entails n tasks (where n is the total number of diseases known to the system) and permits as many as n positive conclusions.

Problem solving procedures for making true/false judgments concerning evoked hypotheses may be of many forms. As already noted, a method commonly employed is that of statistical decision theory. An interesting variant is that of MYCIN, which employs an adaptation of the theorem proving methods of AI to "prove" (up to a specified certainty level) the truth or falsity of each considered hypothesis.

The main difficulty with the binary choice approach is that it fails to aggregate diagnostic possibilities into decision sets; instead, each considered diagnosis is evaluated as though independent of all other alternatives. This requires absolute criteria for decision making, as the problem solver is denied access to powerful heuristics that enable decisions to be rendered relative to a postulated decision set. These heuristics, which provide guidance to the physician concerning the need for additional discriminating information and permit use of efficient decision strategies such as "the process of elimination," can be used only in the context of tasks structured as differential diagnoses.

A second approach that can be used to structure diagnostic decision problems is that used in INTERNIST/I, which is discussed and illustrated in the following section.

6. The INTERNIST–I Model of Diagnostic Reasoning

The INTERNIST–I program employs a heuristic procedure that composes differential diagnoses, dynamically, on the basis of clinical evidence. It does this by assembling what in context appears to be an exhaustive and mutually exclusive subset of disease entities that can explain some significant portion of the observed manifestations of disease. Such a conjectured differential diagnosis then serves as the basis for selecting strategies of information acquisition and decision making relative to that diagnostic task. During the course of an INTERNIST–I consultation, it is not uncommon for a number of such conjectured problem foci to be proposed and investigated, with occasional major shifts taking place in the program's conceptualization of the task at hand.

This system was demonstrated for the first time in 1974, and has since been used in the analysis of hundreds of difficult clinical problems, often with notable success. A review of the information structures and heuristic processes that underlie the performance achieved by this version of INTERNIST follows. Behavior of the system will be illustrated by means of a case run.

7. The INTERNIST–I View of Medical Knowledge

The knowledge base underlying the INTERNIST system is composed of two basic types of elements: disease entities and manifestations (history items, symptoms, physical signs, laboratory data). In addition, there are a number of relations defined on these classes of elements. At present, there are over five hundred disease entities encoded in the knowledge base and thirty-five hundred manifestations.

Each manifestation defines an elementary differential diagnostic task by means of the EVOKES data structure. This structure records with each manifestation the list of diseases in which that manifestation is known to occur, along with a weighting factor (on a 0–5 scale) intended to reflect the strength of association. We refer to this weight as the "evoking strength" by which a manifestation is related to each disease on its "evokes-list."

Although not strictly interpretable as a measure of probability, for reasons that will be discussed presently, this weighting factor can be viewed as somewhat analogous to a posterior probability because it describes the relative likelihood of disease entities evoked on the basis of a single observation. As suggested by this analogy, the ordering of a differential diagnosis provided by evoking strength tends to reflect the a priori probability or prevalence of disease in the population.

The inverse of the EVOKES relation is also recorded explicitly in the knowledge base. By means of the MANIFESTS data structure, each disease entity is profiled with an associated list of manifestations known to occur in that disease, recorded along with an estimate (in this case recorded on a scale of 1–5) of the frequency of occurrence. Although it is not employed in the usual way to calculate posterior probabilities (the actual calculus used in combining these weights will be sketched in the following section), this weighting factor is strictly analogous to the conditional probability of a manifestation given a disease.[1] Other relations are defined on the set of disease entities to record the causal, temporal, and other patterns of association by which the various disease entities and distinguished pathological states are interrelated; these relations also incorporate weighting factors analogous to the evoking strength and frequency weights mentioned above.

The INTERNIST–I knowledge base also contains a nosologic hierarchy of disease categories, organized primarily around the concept of organ systems, having at the top level such categories as "liver disease," "lung disease," "kidney disease," etc. Each of these areas is divided into more specific categories, which may in turn be further subdivided any number of times until the terminal level representing individual disease entities is reached.

There are also several auxiliary relations defined on the class of manifestations to record properties of interest, such as the type of a manifestation (history, symptom, sign, or one of three designations of laboratory finding that reflect progressively greater cost and/or danger to the patient), its global importance (a measure of how important it is to account for that manifestation in a final diagnosis, recorded on a scale of 1–5), and its relation to other manifestations (such as the derivability of one from another).

8. The Problem-formation Method of INTERNIST-I

The heuristic problem structuring procedure of INTERNIST–I is invoked repeatedly during the course of a diagnostic consultation in order to deal sequentially with the component parts of a complex clinical problem.

---

[1] Unlike the evoking strength, which tends to be a subjective estimate that requires extensive clinical experience, the frequency weights can often be supported on the basis of somewhat more objective data, obtained from a careful review of the literature. Still, these estimates must be characterized as largely subjective, due to the high degree of variability in the literature.

The process is as follows. First, during the initial input phase, patient data consisting of both positive findings and pertinent negative findings may be entered in any order and any amount. Each positive finding is used to evoke an elementary differential diagnostic task that may contain a mixture of individual disease nodes and higher level nodes of the disease hierarchy. Where a manifestation evokes one of the higher level category nodes, it is because that finding can be explained by all subnodes of the given category.

In order to detect the possibility of combining multiple tasks, each associated with a single abnormal finding, the INTERNIST–I program uses a scoring procedure that awards credit to disease hypotheses on the basis of the number and the importance of elementary tasks that are unified by those hypotheses. In this scoring process, the evoking strength and importance of manifestations explained by a disease are counted in its favor; frequency weights count against those disease hypotheses in which the corresponding manifestations are expected but not found present in the case.[2]

Given a ranked list of disease hypotheses, a synthesized task definition is then formulated on the basis of the most highly rated of these items, using the following heuristic criterion: two disease entities are considered to be alternatives to one another (hence part of the same task definition) if, taken together, they explain no more of the observed findings than are explained by one or the other separately. The set of alternatives so determined, with scores within a fixed range of the top ranked disease hypothesis on the list, are then composed into a differential diagnostic task, which becomes the focus of problem-solving attention.

The program then selects questions that will help to discriminate among entities in the problem set. There are three major strategies that may be employed in selecting questions to be asked and identifying tests to be performed. When the set of decision alternatives contains five or more elements, a RULEOUT strategy is invoked, whereby the program seeks negative findings that will help to disconfirm one or more elements in the set. If the decision set contains between two and four elements, the program fixates on the leading two contenders and seeks information that will help separate them in score; this is referred to as a DISCRIMINATE information acquisition strategy. When only one alternative remains in the decision

---

[2] The scoring process employs a simple geometric mapping, used to reflect the intention of the knowledge base designers in assigning weights; namely, that two manifestations with an evoking strength of four would be considered equivalent to one with an evoking strength of five, two with evoking strength of three are equivalent to one of four, etc. Similar considerations apply in the assignment of the frequency and importance weighting factors; hence, they, too, are subjected to a geometric transformation before being combined algebraically as outlined above to obtain the total score assigned to each evoked disease hypothesis.

set, the program embarks on a PURSUING strategy, whereby it seeks confirmatory data that will extend the separation of the leading contender from its nearest competitor above a given threshold value.

Once the program has inquired about the selected information items and acquired additional positive or negative patient data, it then re-evaluates all diseases evoked (whether in or out of the current problem focus) on the basis of new information obtained, and then reformulates the differential diagnosis. Depending on which disease entity emerges as most highly rated on successive iterations of the process, the focus of attention may shift from one diagnostic task to another—but at any one time, there is a single problem under active consideration.

Whenever a diagnostic task becomes solved, its result is entered into a list of concluded diagnoses; all manifestations explained by that disease are marked "accounted for"; and the process recycles until all problems present in the case have been uncovered and solved. Because of causal, temporal, or other interrelationships, certain combinations of disease entities are more likely to occur than others. This fact is recognized in INTERNIST–I by the scoring algorithm, which on each iteration of the process, gives additional weight to any disease entity that is in any way linked to some already concluded disease.

In the course of gathering additional information, it sometimes happens that the program runs out of questions deemed useful for the strategy being pursued. If this happens while engaged in a PURSUING strategy, the program renders a peremptory conclusion even though the desired threshold of separation has not been achieved. If the program exhausts its list of useful questions during a DISCRIMINATE or RULEOUT phase, however, it performs what we refer to as a "deferral." At this point, the set of findings used to define the current differential diagnostic task are set aside for the time being (as though a positive decision had been made that some unknown member of that decision set is actually correct) so that some other decision problem might be brought into consideration. The hope, often borne out in practice, is that by solving a second — possibly related — decision problem, the program might then go back to the deferred differential diagnostic task and make somewhat more progress in its resolution.

By virtue of this deferral mechanism, INTERNIST–I exhibits a limited capability for concurrent problem formation and problem solving, an important characteristic of human diagnostic behavior.

9. Example of INTERNIST–I Case Analysis

The following is a transcript of the interaction that took place during a recent INTERNIST–I case study. The data were taken from records of a patient admitted to the hospital with a severe acute febrile illness, which was correctly diagnosed by the house staff as systemic leptospirosis. Not a particularly challenging diagnostic problem, this case was selected for demonstration purposes because it illustrates a number of important aspects of the INTERNIST–I approach.

(DOCTOR)

This command is used to invoke the DOCTOR program, which embodies the interactive diagnostic procedure of INTERNIST–I; the program is written in a dialect of LISP called INTERLISP.

INTERNIST–I consultation SUMEX-AIM version
PLEASE ENTER FINDINGS.

After the initial prompt PLEASE ENTER FINDINGS, the person using the system enters whatever positive findings he chooses to divulge. The program prompts with "*" to request additional input until the respondent types "GO"; after which INTERNIST–I commences the task definition procedure described in the previous section. Had the user wanted to enter negative findings during this initial exchange, he could have responded "DENY" to the prompt, causing subsequent entires to be interpreted as negative results.

```
*SEX MALE
*AGE 26 TO 55
*RACE WHITE
*ALCOHOLISM CHRONIC HX
*EXPOSURE TO RABBITS OR OTHER SMALL MAMMALS
*FEVER
*MYALGIA
*LEG <S> WEAKNESS BILATERAL
*LEG <S> WEAKNESS PROXIMAL ONLY
*PRESSURE ARTERIAL SYSTOLIC 90 TO 110
*PRESSURE ARTERIAL ORTHOSTATIC HYPOTENSION
*TACHYCARDIA
*JAUNDICE
*ANOREXIA
*DIARRHEA ACUTE
*FECES LIGHT COLORED
*VOMITING RECENT
*LIVER ENLARGED SLIGHT
*SKIN SPIDER ANGIOMATA
*SKIN PALMAR ERYTHEMA
*OLIGURIA HX
*UREA NITROGEN BLOOD 60 TO 100
*CREATININE BLOOD INCREASED
*WBC 14000 TO 30000
*PLATELETS LESS THAN 50000
*GO
```

Manifestations are expressed by means of precise sequences of terms in a controlled vocabulary; there are presently approximately 3500 vocabulary items that can be used to describe positive and negative findings.

> DISREGARDING: EXPOSURE TO RABBITS OR OTHER SMALL MAMMALS, LEG <S> WEAKNESS BILATERAL, LEG <S> WEAKNESS PROXIMAL ONLY, PRESSURE ARTERIAL ORTHOSTATIC HYPOTENSION, CREATININE BLOOD INCREASED, UREA NITROGEN BLOOD 60 TO 100
>
> CONSIDERING: AGE 26 TO 55, SEX MALE, ANOREXIA, MYALGIA, VOMITING RECENT, FECES LIGHT COLORED, FEVER, JAUNDICE, LIVER ENLARGED SLIGHT, SKIN PALMAR ERYTHEMA, SKIN SPIDER ANGIOMATA, WBC 14000 TO 30000, PLATELETS LESS THAN 50000
>
> RULEOUT: HEPATITIS CHRONIC ACTIVE, ALCOHOLIC HEPATITIS, HEPATIC MILIARY TUBERCULOSIS, MICRONODAL CIRRHOSIS <LAENNECS>, HEPATITIS ACUTE VIRAL

At this point, INTERNIST–I reports concerning the initial differential diagnosis that will be the focus of problem-solving attention. Three lists are displayed, labelled respectively DISREGARDING, CONSIDERING, and RULE-OUT. The CONSIDERING list identifies those positive findings whose differential diagnostic tasks were combined in coming to the differential diagnostic focus, which is described by the RULEOUT list. The DISREGARDING list tells what positive findings are not consistent with the differential diagnosis as formulated and are therefore being disregarded for the moment; they will, however, be attended to in due course. The keyword RULEOUT indicates what strategy INTERNIST–I is going to pursue relative to this task definition: in this case, as in any case when the differential list contains five or more alternatives, the program will try to find questions to ask which, if the response is negative, will serve to downgrade the score of one or more diagnoses on the list. While pursuing a RULEOUT strategy, the program limits the questions to be asked to items of information obtainable via history or physical exam.

> Please Enter Findings of JAUNDICE HX
> *GO

There are two formats used for asking questions. In the preceding line, the user is asked to provide any data that might be available within the specified category of findings. Actually, the respondent is free to enter whatever positive or negative data desired and is not constrained to the category mentioned in the query. In this case, the user chose to respond GO; this passes the initiative back to the program, which then typically follows up on the general question by asking about the specific finding of that category for which it particularly wants a YES or NO answer.

> JAUNDICE INTERMITTENT HX?
> NO

If the respondent did not have any information concerning this direct question, he could have answered N/A—meaning not available.

```
Please Enter Findings of APPETITE/WEIGHT
*GO
WEIGHT LOSS GTR THAN 10 PERCENT?
NO
```

The program asks a series of questions that have been selected in accordance with its problem-solving strategy, then repeats the scoring and partitioning of the task definition procedure.

```
    DISREGARDING: EXPOSURE TO RABBITS OR OTHER SMALL MAMMALS, LEG <S>
WEAKNESS BILATERAL, LEG <S> WEAKNESS PROXIMAL ONLY, PRESSURE ARTERIAL
ORTHOSTATIC HYPOTENSION, CREATININE BLOOD INCREASED, UREA NITROGEN
BLOOD 60 TO 100
    CONSIDERING: AGE 26 TO 55, SEX MALE, ANOREXIA, MYALGIA, VOMITING
RECENT, FECES LIGHT COLORED, FEVER, JAUNDICE, LIVER ENLARGED SLIGHT, SKIN
PALMAR ERYTHEMA, SKIN SPIDER ANGIOMATA, WBC 14000 TO 30000, PLATELETS
LESS THAN 50000
    RULEOUT: HEPATITIS CHRONIC ACTIVE, ALCOHOLIC HEPATITIS, HEPATIC MILIARY
TUBERCULOSIS, HEPATITIS ACUTE VIRAL, INFECTIOUS MONONUCLEOSIS
```

Except for the substitution of an acute process (infectious mononucleosis) for a chronic one (micronodal cirrhosis), this differential diagnosis is not significantly changed from the initial formulation. Note that the possibility of cirrhosis has not actually been ruled out; it has merely dropped out of sight because its score has fallen below the threshold used by the task definition procedure.

```
Please Enter Findings of PAIN ABDOMEN
*GO
ABDOMEN PAIN GENERALIZED?
NO
ABDOMEN PAIN EPIGASTRIUM?
NO
ABDOMEN PAIN NON COLICKY?
NO
ABDOMEN PAIN RIGHT UPPER QUADRANT?
NO
    DISREGARDING: JAUNDICE, SKIN SPIDER ANGIOMATA, CREATININE BLOOD
INCREASED, UREA NITROGEN BLOOD 60 TO 100
    CONSIDERING: AGE 26 TO 55, EXPOSURE TO RABBITS OR OTHER SMALL
MAMMALS, SEX MALE, ANOREXIA, DIARRHEA ACUTE, MYALGIA, VOMITING RECENT,
FEVER, LEG <S> WEAKNESS BILATERAL, LEG <S> WEAKNESS PROXIMAL ONLY,
```

PRESSURE ARTERIAL ORTHOSTATIC HYPOTENSION, PRESSURE ARTERIAL SYSTOLIC
90 TO 110, TACHYCARDIA, WBC 14000 TO 30000, PLATELETS LESS THAN 50000
DISCRIMINATE: LEPTOSPIROSIS SYSTEMIC, SARCOIDOSIS CHRONIC SYSTEMIC

The effect of the negative responses concerning abdominal pain has been
to lower the scores of all of the hepatic disorders considered in the previous
differential diagnosis. This time, when the partitioning algorithm is invoked
the highest-ranking alternative is systemic leptospirosis; the only other
diagnosis on the list capable of explaining substantially the same set of
findings is systemic sarcoidosis. The keyword DISCRIMINATE indicates that
the list of alternatives contains between two and four elements, the leading
two of which are selected for comparative analysis. When engaged in a
DISCRIMINATE mode of analysis, the program will attempt to ask questions
serving to support one diagnosis at the expense of the other; more costly
procedures may be called for in order to achieve this objective.

        Please Enter Findings of VOMITING/REGURGITATION
        *GO

        HEMATEMESIS?
        NO

        HEMOPTYSIS GROSS?
        NO

        Please Enter Findings of TEMPERATURE
        *GO

        RIGOR <S> ?
        YES

        Please Enter Findings of NEUROLOGIC EXAM CRANIAL NERVE <S>
        *GO

        NERVE PARALYSIS SEVENTH CRANIAL BILATERAL?
        NO

        SPLENECTOMY HX?
        NO

The program is not actually interested in the answer to this question; what
it wants to know is whether the spleen is enlarged. Because of the possibility
of being misled by a negative answer, appropriate blocks have been created
to prevent the program from asking about an increased spleen size in a
patient whose spleen had been removed.

        Please Enter Findings of PALPATION ABDOMEN
        *GO

        SPLENOMEGALY MODERATE?
        NO

        Please Enter Findings of XRAY LUNG FIELD <S>
        *GO

CHEST XRAY HILAR ADENOPATHY BILATERAL?
NO

DISREGARDING: JAUNDICE, SKIN SPIDER ANGIOMATA, CREATININE BLOOD INCREASED, UREA NITROGEN BLOOD 60 TO 100

CONSIDERING: AGE 26 TO 55, EXPOSURE TO RABBITS OR OTHER SMALL MAMMALS, SEX MALE, ANOREXIA, DIARRHEA ACUTE, MYALGIA, VOMITING RECENT, FEVER, LEG <S> WEAKNESS BILATERAL, LEG <S> WEAKNESS PROXIMAL ONLY, PRESSURE ARTERIAL ORTHOSTATIC HYPOTENSION, PRESSURE ARTERIAL SYSTOLIC 90 TO 110, RIGOR <S>, TACHYCARDIA, WBC 14000 TO 30000, PLATELETS LESS THAN 50000

PURSUING: LEPTOSPIROSIS SYSTEMIC

The questions about rigors (shaking chills) is enough to separate the scores of these items so that now there is only one alternative left in the differential diagnosis. Before concluding that this diagnosis is actually correct, however, the program will now attempt to achieve a degree of separation between this diagnosis and its nearest competitor (now below the threshold and not printed out) that is twice the threshold value. The program invokes a PURSUING strategy, which calls for the identification and acquisition of clinching data; at this stage, the level of questioning is unconstrained so the program can ask about biopsies if useful, or other specialized procedures capable of providing pathognomonic data.

LEPTOSPIRA AGGLUTINATION POSITIVE?
YES

This finding is enough to clinch the diagnosis. However, the program proceeds to ask the additional questions that had been prepared for this round of information acquisition. As implemented in this program, the process of task formulation is too time consuming to have the procedure recycle after each new datum is entered.

Please Enter Findings of BLOOD CULTURE
*GO

BLOOD CULTURE LEPTOSPIRA?
N/A

Please enter Findings of URINE CULTURE
*GO

URINE CULTURE LEPTOSPIRA?
N/A

DISREGARDING: JAUNDICE, SKIN SPIDER ANGIOMATA, CREATININE BLOOD INCREASED, UREA NITROGEN BLOOD 60 TO 100

CONSIDERING: AGE 26 TO 55, EXPOSURE TO RABBITS OR OTHER SMALL MAMMALS, SEX MALE, ANOREXIA, DIARRHEA ACUTE, MYALGIA, VOMITING RECENT, FEVER, LEG <S> WEAKNESS BILATERAL, LEG <S> WEAKNESS PROXIMAL ONLY, PRESSURE ARTERIAL ORTHOSTATIC HYPOTENSION, PRESSURE ARTERIAL SYSTOLIC

90 TO 110, RIGOR <S>, TACHYCARDIA, WBC 14000 TO 30000, LEPTOSPIRA
AGGLUTINATION POSITIVE, PLATELETS LESS THAN 50000
    PURSUING: LEPTOSPIROSIS SYSTEMIC
    CONCLUDE: LEPTOSPIROSIS SYSTEMIC

Having concluded leptospirosis, the program now deletes from further
consideration those items explained by this diagnosis. If there are significant
data remaining to be explained, the task formulation procedure is again
invoked to determine the next most compelling differential diagnosis. In
the scoring process, any diagnoses that are causally linked to the concluded
disease are awarded an extra bonus. This enhances the likelihood of their
being noticed by the task formulator, but does not guarantee that they will
be considered.

    DISREGARDING: CREATININE BLOOD INCREASED, UREA NITROGEN BLOOD 60 TO
100
    CONSIDERING: ALCOHOLISM CHRONIC HX, FECES LIGHT COLORED, JAUNDICE,
LIVER ENLARGED SLIGHT, SKIN PALMAR ERYTHEMA, SKIN SPIDER ANGIOMATA
    RULEOUT: MICRONODAL CIRRHOSIS <LAENNECS>, HEPATIC LEPTOSPIROSIS,
FATTY LIVER SECONDARY, CIRRHOSIS SECONDARY TO CHOLESTATIC DISEASE,
ALCOHOLIC HEPATITIS, MACRONODAL CIRRHOSIS <POSTNECROTIC>, DRUG
HYPERSENSITIVITY CHOLESTATIC REACTION

As the signs of acute febrile illness have been accounted for by the conclusion of systemic leptospirosis, the diagnostic task formulated to deal
with the liver involvement is now focused primarily on those chronic disorders that commonly cause the type of skin lesions reported in this patient.
Note, however, that the possibility of hepatic involvement by leptospirosis—
never mentioned in the earlier differential diagnoses dealing with the liver
involvement—is now high on the list of alternatives.

    LIVER EDGE HARD?
    N/A

    ONSET ABRUPT?
    YES

    JAUNDICE CHRONIC PERSISTENT HX?
    NO

    Please Enter Findings of INSPECTION HAND <S> AND FEET
    *GO

    FINGER <S> CLUBBED?
    NO

    Please Enter Findings of INSPECTION AND PALPATION SKIN
    *GO

    SKIN RASH MACULOPAPULAR?
    NO

LIVER DISTORTED OR ASYMMETRICAL?
N/A

Please Enter Findings of INSPECTION AND PALPATION EXTREMITY <IES>
*GO

LEG <S> EDEMA BILATERAL?
NO

ABDOMEN TENDERNESS GENERALIZED?
NO

ABDOMEN TENDERNESS RIGHT UPPER QUADRANT?
NO

DISREGARDING: SKIN SPIDER ANGIOMATA, CREATININE BLOOD INCREASED, UREA NITROGEN BLOOD 60 TO 100

CONSIDERING: ONSET ABRUPT, FECES LIGHT COLORED, JAUNDICE, LIVER ENLARGED SLIGHT

DISCRIMINATE: HEPATIC LEPTOSPIROSIS, LARGE DUCT OBSTRUCTION

The abrupt onset of the illness tends to eliminate the chronic liver diseases from serious contention. However, the program does not yet have access to the results of liver function tests, which would help to discriminate between hepatocellular and cholestatic forms of acute involvement.

Please Enter Findings of TRANSAMINASE <S>
*SGOT GTR THAN 400

*SGPT 200 TO 600

*GO

DISREGARDING: SKIN SPIDER ANGIOMATA, CREATININE BLOOD INCREASED, UREA NITROGEN BLOOD 60 TO 100

CONSIDERING: FECES LIGHT COLORED, JAUNDICE, LIVER ENLARGED SLIGHT, SGOT GTR THAN 400, SGPT 200 TO 600

NARROW: HEPATIC LEPTOSPIROSIS, HEPATITIS ACUTE VIRAL

The term NARROW signifies that the task formulation procedure has constructed a broad differential list that would ordinarily suggest use of the RULEOUT strategy; however, as there are no helpful questions available without resorting to laboratory procedures, the field is arbitrarily restricted to the two leading contenders, and the DISCRIMINATE strategy is employed.

Please Enter Findings of BLOOD COUNT AND SMEAR
*GO

WBC GTR THAN 30000?
YES

Please Enter Findings of BLOOD PHOSPHATASE <S>
*GO

ALKALINE PHOSPHATASE BLOOD GTR THAN 2 TIMES NORMAL?
YES

ABDOMEN PAIN EXACERBATION WITH MEAL <S>?
NO

ABDOMEN GUARDING INVOLUNTARY GENERALIZED?
NO

ABDOMEN GUARDING INVOLUNTARY LOCALIZED?
NO

Please Enter Findings of TRANSFUSION <S> HX
*GO

TRANSFUSION <S> MULTIPLE HX?
NO

DISREGARDING: SKIN SPIDER ANGIOMATA, CREATININE BLOOD INCREASED,
UREA NITROGEN BLOOD 60 TO 100

CONSIDERING: FECES LIGHT COLORED, JAUNDICE, LIVER ENLARGED SLIGHT,
ALKALINE PHOSPHATASE BLOOD GTR THAN 2 TIMES NORMAL, SGOT GTR THAN 400,
WBC GTR THAN 30000, SGPT 200 TO 600

PURSUING: HEPATIC LEPTOSPIROSIS
CONCLUDE: HEPATIC LEPTOSPIROSIS

DISREGARDING: SKIN SPIDER ANGIOMATA, UREA NITROGEN BLOOD 60 TO 100

CONSIDERING: OLIGURIA HX, CREATININE BLOOD INCREASED

RULEOUT: RENAL LEPTOSPIROSIS, GLOMERULONEPHRITIS ADVANCED CHRONIC,
LEAD NEPHROPATHY, GLOMERULONEPHRITIS ACUTE, PYELONEPHRITIS CHRONIC,
DIABETIC NEPHROPATHY

Having decided the nature of the hepatic involvement, INTERNIST–I now turns its attention to the differential diagnosis of renal disease in this patient. Although this problem was explored extensively by the program via a score or more additional questions, no conclusion was ever reached; the final statement of this differential at the time the program was interrupted appeared essentially as it does in this display.

10. References

Buchanan, B. G., & Feigenbaum, E. A. DENDRAL and meta-DENDRAL: Their applications dimension. *Artificial Intelligence,* 1978, *11,* 5–24.

Duda, R. O., Hart, P. E., Nilsson, N. J., & Sutherland, G. L. Semantic network representations in rule-based inference systems. In D. A. Waterman & F. Hayes-Roth (Eds.), *Pattern-directed inference systems.* New York: Academic Press, 1978.

Miller, R. A., Pople, H. E., & Myers, J. D. INTERNIST–I, An experimental computer-based diagnostic consultant for general internal medicine. *New England Journal of Medicine,* August 19, 1982, *307,* 468–476.

Pople, H. E. Heuristic methods for imposing structure on ill structured problems: The structuring of medical diagnostics. In P. Szolovits (Ed.), *Artificial intelligence in medicine.* Boulder, CO: Westview Press, 1982.

Shortliffe, E. H. *Computer based medical consultations: MYCIN.* New York: American Elsevier, 1976.

Expert Systems: Matching Techniques to Tasks*

B. *Chandrasekaran*
DEPARTMENT OF COMPUTER
AND INFORMATION SCIENCE
OHIO STATE UNIVERSITY
COLUMBUS, OH 43210

In this chapter an attempt is made to relate the architectures and representations for expert sysems to the types of tasks for which they are appropriate. We start with an analysis of the features that characterize an expert system, and discuss the need for symbolic knowledge structures that support qualitative reasoning in the design of expert systems. We consider rules, logical formulas and frames for representation of expert knowledge. In particular we provide an analysis of the multiplicity of roles that rules have played in different rule-based systems, and emphasize the need to distinguish among these roles. We proceed to outline our theory of types of expert problem solving and argue that such a taxonomy enables one to characterize expert system capabilities and help match problems with techniques. Throughout the chapter it is emphasized that the important issue is the nature of the information processing task in a given task domain, and issues of formalisms for representation are subordinate to that basic issue.

1. Expert Systems: What Are They?

It is clear that recently artificial intelligence has created enormous excitement in the commercial marketplace, and one of the sources of this excitement is the promise of expert systems or knowledge-based systems in solving or assisting in the solution of many practical problems. There is a widespread feeling that *knowledge* is the next frontier in the practical application of computers, and the well-publicized commitment of the Japanese to research on and development of a Fifth Generation computer for knowl-

---

* Research supported by AFOSR Grant 82–0255.

edge processing has only added to this sense of an impending revolution. The phrase "expert systems" evokes all sorts of hopes: From the clerical worker to the research scientist in a corporation, each employee is a storehouse of an enormous amount of knowledge and problem-solving capabilities. The syllogism goes something like, "I need an expert for X, they are hard to find or expensive, thus I need an expert system for X." It is clear to most of us doing research in the field that while the promise of carefully deployed expert systems is indeed high, we are nowhere near the stage where we can hope to replace all kinds of human experts with computer-based expert systems. To take an extreme example, advanced researchers or creative mathematicians clearly perform knowledge-based expert problem solving, but equally clearly the state of the art in expert systems is not up to replacing them with expert systems. Any attempt to characterize, even if informally, what sorts of problems are amenable to current techniques — matching techniques to tasks, if you will — could be very useful.

A related problem is that, in strict definitional terms, it is hard to be very precise about what characteristics of a computer program to solve a class of problem qualify it to be called an expert system. The following dimensions have often been suggested as important in such a definition.

- *Expertise.* Certainly a necessary condition is that an expert system should have expertise in the domain and show expert-level performance is some aspects of the domain. However, this condition is not sufficient. Is a payroll program written in COBOL an expert system? In some real sense it captures the expertise of an accountant whose domain knowledge is incorporated in the many branching decisions made by the logic of the program.
- *Search.* The intuition that a program must do some search in a space of possibilities — following the idea that search is an essential characteristic of intelligent reasoning — is generally useful but not always valid, because R1 (McDermott, 1982), an expert system that has been very successful in practical use, does not do any search in the execution of its main task.
- *Uncertainty.* While uncertainty in data or knowledge gives many expert problem domains (such as medicine) interesting additional properties and makes them challenging for the designer of expert systems, it is not a defining characteristic, since, again using R1 as an example, its knowledge and data do not involve probabilistic or other types of uncertainty.
- *Symbolic knowledge structures.* Most expert systems in the AI field have their domain knowledge in explicitly symbolic form as collec-

tions of facts, rules, frames etc., which are explicitly manipulated by problem solving or inference mechanisms to produce answers to questions. This is to be contrasted with mathematical or simulation models or mainly numerical information as the main knowledge base. However, many well-known expert systems such as INTERNIST represent the bulk of their core knowledge in the form of numerical relations between entities. As an incidental observation, we may note that the general emphasis in expert systems work on symbolic knowledge structures often elicits another sort of response. Engineers trained in mathematical modelling techniques find it difficult to understand why a computer program which evaluates, say, a system of complex mathematical equations describing some process (such as nuclear reaction in a reactor vessel) is not an expert system — after all, they argue, such a system is an embodiment of very highly specialized expert knowledge, capable of providing answers to a number of questions. Further there is a tendency among some people in this group to regard AI programs as "approximate," because the rules that an expert system will use are supposed to be only heuristic, while the mathematical models are exact. (See section 3.2. for a discussion of this issue.)

- *Explanation capability.* Continuing in the vein of searching for definitional characteristics of expert systems, the idea that such systems should be able to *explain* their reasoning is a useful constraint on the structure and functioning of expert systems, but as a rule, since the activity of explanation itself is poorly understood, it is not yet certain what structures and functions this requirement rules out or permits.

- *Other features.* Many people in the industrial world have virtually taken to *defining* expert systems as those that have a knowledge base and a separate inference engine, and that knowledge base in expert systems should necessarily be in the form of *rules.* Again, while rules have been a dominant method for knowledge representation in many first generation expert systems, many expert systems have used frame structures (Pauker, Gorry, Kassirer, & Schwartz, 1976), or network structures (Duda, Gaschnig, & Hart, 1979). In any case, this emphasis on knowledge representation formalisms often obscures the more fundamental issues of the *content* of the computation these systems do. For example, Szolovits and Pauker (1978) point out that MYCIN, the system most widely thought of as a prototypical example of the rule-based approach can be recast as a frame-based system which uses procedural attachments to fill the slots in various frames.[1] Taking up the knowledge base/inference engine separation issue, it is unlikely that a complete separation of knowledge and inference is viable as a basic principle in the organization of expert systems. We (Chandrasekaran, 1982; Gomez and Chandrasekaran, 1981) as well as, more recently, others (Davis, 1982; Stefik, et al., 1982) have argued that knowledge and its use are likely to be more strongly intertwined as the difficulties and variety of the tasks the expert systems are called upon to perform increase. (This argument will be elaborated in section 7 of this paper.)

The point of the foregoing is not to suggest a precise, issue-settling definition of expert systems, but merely to point to the multiple dimensions along which expert systems can be viewed, and to the need for more careful analysis of much of the terminology that is used in discussing expert systems.

The major line of argument that we will pursue in this paper can be outlined as follows. In section 2, we briefly trace the development of the idea of knowledge-based systems in AI. Section 3 is devoted to discussing the increasing need for symbolic content to expert reasoning as the size and demands of the task domain increase; i.e., we will analyze why a complete mathematical model of the situation, even if available, will not meet many of the demands placed on expert reasoning. In section 4, we discuss the several distinct senses and roles that rules can play and have played in expert systems, and how a failure to keep these separate can cause a great deal of confusion. In section 5, we briefly discuss logic-based and frame-based representations. In section 6, we discuss the need for *organization* of knowledge for effective use, and in section 7, we argue that further organizational constructs, such as *concepts* and *types of problem solving,* are needed both to construct more powerful expert systems, and to characterize their capabilities. This will also have the side effect of emphasizing the increasing need not to separate knowledge bases and inference machineries. We will also provide in this section two examples of *generic* problem-solving types, and show how each type of problem solving

---

[1] There is a general problem in AI of not making clean distinctions between the *basic information processing task* of a computation and the algorithm or program that carries out this task. Marr (1976) presents arguments for such a distinction in computational theories of vision. In Gomez and Chandrasekaran (1981) we make analogous arguments in the area of knowledge representation for problem solving systems. Saying that System A uses rules, while System B uses networks for knowledge representation says nothing about the nature of the information processing activities that go on in the two systems: a comparison at the formalism level would miss many important aspects of the similarities and differences that ought to be sought at a higher level. Our discussions later in the paper regarding rules elaborate on some aspects of this point.

induces an organization of knowledge in the form of a cooperating community of "specialists" engaged in that problem-solving type. At this point, we will be able to discuss what sorts of problems can be handled by "compiled" structures and what sorts need "deeper" problem solving. In that context we will discuss the issues related to the so-called *causal modeling* problem. This will help us present a discussion of the issues surrounding the degree of *understanding* that expert systems may need to have about the domain. The overall flow of the discussion is in the direction of the evolution of expert systems from numerical programs to highly organized symbolic structures engaged in distinct types of problem solving and communicating with one another.

An admission is in order at this point. The title is more ambitious than what we will be able to accomplish in this paper. A really satisfactory account must await a better overall theory of problem solving than we have at present. Even an imcomplete theory, however, such as the one presented in section 7, is capable of providing a framework for characterizing capabilities of expert systems in generic terms. Thus the paper should be viewed as stating a position on the sorts of theories that might help us characterize in powerful ways how the "engineering" of knowledge might rest on a more systematic understanding.

2. On the Idea of Knowledge-Based Systems

A historic reminder may be useful to clarify the term, "knowledge-based." This phrase, in the context of AI systems, arose in response to a recognition, mainly in the pioneering work of Feigenbaum and his associates in the early to mid–1970s, that much of the power of experts in problem solving in their domains arose from a large number of rule-like pieces of domain knowledge which helped constrain the problem solving effort and direct it towards potentially useful intermediate hypotheses. These pieces of knowledge were domain-specific in the sense that they were not couched in terms of general principles or heuristics which could be instantiated within each domain, but rather were directly in the form of appropriate actions to try in various situations. The situations corresponded to partial descriptions stated in terms of domain features. This hypothesis about the source of expert problem-solving power was in contrast to the previous emphasis in problem-solving research on powerful *general* principles of reasoning which would work on different domain representations to produce solutions for each domain. Thus the *means-ends heuristic* of the General Problem Solver program (Newell & Simon, 1972) is a general-purpose heuristic, which would attempt to produce solutions for different problems by working on the respective problem representations. On the other hand, a piece of knowledge like, "When considering liver diseases, if the patient has been exposed to certain chemicals, consider hepatitis," is a domain-specific heuristic which the human expert was said to use directly.

The paradigm for knowledge-based systems that was elaborated consisted of extracting from the human experts a large number of such rules for each domain and creating a knowledge-base with such rules. It was generally assumed as part of the paradigm that the rule-using (i.e., reasoning) machinery, was not the source of problem-solving power, but rather the rules in the knowledge base. Hence the slogan, "In Knowledge Lies the Power."

The word "knowledge" in "knowledge base systems" is used in a rather special sense. It is meant to refer to the knowledge an expert problem solver in a domain was posited to have which gives a great deal of efficiency to the problem-solving effort. The alternative against which this position is staked is one in which complex problem solving machineries are posited to operate on a combination of basic knowledge that defines the domain with various forms of general and common sense knowledge. Thus the underlying premise behind much of the AI work on expert systems is that once the body of expertise is built up, expert reasoning can proceed without any need to invoke the general world and common sense knowledge structures. If such a decomposition were not in principle possible, then the development of expert systems would have to await the solution of the more general problem of common sense reasoning and general world knowledge structures. What portion of expert problem solving in a given domain can be captured in this manner is an empirical question, but experience indicates that it is at least a nontrivial subset in various important domains. This decoupling of common sense and general purpose reasoning from domain expertise is also the explanation for the rather paradoxical situation in AI where we have programs which display, e.g., expert-like medical diagnostic capabilities while the field is still some distance from capturing most of the intellectual activities that children do with ease.

We will argue later in the paper that while many first-generation expert systems have been successful in relatively simple problems with this approach, which lays relatively greater emphasis on heuristic knowledge specific to the domain at the expense of the problem solving aspects, the next generation of research in this area — as well as application systems — will be bringing back an increased emphasis on the latter.

3. From Numerical to Symbolic Modeling of Expertise

In this section we will provide two sets of reasons why symbolic structures become necessary to support decision making. By considering the example of multivariate prediction, we will suggest that certain kinds of computational problems are alleviated by multiple-level symbolic structures. In the next subsection, we will argue that certain kinds of decisions cannot be made purely within a numerical model, however complete it may be in principle.

3.1. Multivariate Classification

There is an ubiquitous but conceptually simple class of problems in decision making which can often be typically modeled as mapping a multivariate state vector to a set of discrete categorical states. The classical pattern classification paradigm deals with this class of problems. There are many application domains in which such problems arise locally, and domain experts may be called upon to perform such a classificatory task as part of a larger problem-solving effort. Classifying weather conditions based on a number of measurements, and predicting the presence or absence of certain pathophysiological conditions on the basis of a vector of numerical predictor variables are examples of this task. Often, once the predictor variables themselves are chosen in consultation with the experts, and when the state vector is of modest dimensionality (in the order of 10's), a discriminant type of analysis can be as good as or better than human experts. Much theoretical effort in pattern recognition notwithstanding, experience in this type of task was that this is an example where power came from expert choice of the variables, rather than in the complexity of the discriminants themselves. But when the dimensionality of the state vector gets very large this approach has serious problems both in one's ability to compute the discriminant as well as in the sensitivity of the decision to small changes. Using the medical example earlier, the totality of the medical diagnostic problem can be formally viewed as a mapping from a (very large) vector of manifestations to a (again quite large) set of named diseases. What works very well locally, i.e., for state vectors of low dimensionality and few decision states, deteriorates rapidly when the sizes get large. It doesn't matter which sort of discriminant one uses: statistical ones or perceptron-like threshold devices. The computational problems in the statistical case are described well in Szolovits and Pauker (1978). Corresponding difficulties for perceptron-like devices, especially those relating to sensitivity issues, are discussed by Minsky and Papert (1969).

One approach to overcoming the above computational problems is to introduce multiple *layers* of decisions. Instead of one classification function from say a 200–dimensional state vector, the problem can be broken into groups of (possibly overlapping) state vectors of the order of 10's, each of them providing a small number of discrete values as local mappings. Typ-

ically these groupings will correspond to potentially meaningful intermediate entities, so that one can interpret each grouping as computing a symbolic abstraction corresponding to an intermediate concept. At the second
layer, the outputs of each of these can be grouped together in a similar
manner and the process repeated. This is precisely what Samuel's Signature
Tables (1967) did in transforming a description of a checker board configuration into a classification in terms how good the board was for a player.
Each stage of the abstraction is computationally simple. An important point
to notice is that a shift away from numerical precision towards discretization
and symbolization is taking place here in capturing expertise.

3.2. Formal Models vs Symbolic Knowledge Structures

Even if one had a complete mathematical model of a situation, that by itself
is not often sufficient for many important tasks. All the numerical values
of the various state variables will still need to be *interpreted*. Identifying
interesting and potentially significant states from the initial conditions requires *qualitative* reasoning rather than a complete mathematical simulation. To take a pedagogically effective but fanciful example, consider a
household robot watching its master carelessly move his arm towards the
edge of a table where sits a glass full of wine. In theory a complete mathematical equation of the arm and all the objects in the room including the
volume of wine is possible. But the most detailed numerical solution of
this will still only give values for a number of state variables. It still takes
further reasoning to interpret this series of values and arrive at a simple
common sense level statement, viz., "the arm will hit the wine glass and
wine will spill on the carpet." On the other hand the aim of "naive physics"
models is to support qualitative reasoning that can arrive at such conclusions readily. The reason why a complete numerical solution is not enough,
i.e., why the above symbolic conclusion *ought* to be reached by the robot,
is that its own knowledge of available actions is more appropriately indexed
by symbolic abstractions as "spilled wine," rather than by the numerical
values of all ranges of relevant state variables in the environment.

Thus when faced with reasoning tasks involving complex systems, both
human experts as well as expert systems are necessarily compelled to deal
with symbolic knowledge structures, whether or not complete mathematical
models may in principle be available. Human experts (as well as AI expert
systems) may, at specific points during their qualitative reasoning, switch
to a local formal analysis, such as a medical specialist using formulas or
equations to decide which side of the acid/base balance a patient may be
in, but this formal analysis is under the overall control of a symbolic reasoning system. It is the symbolic knowledge and problem-solving structures
that are of central interest to the science and technology of AI.

4. On the Role of Rules in Rule-Based Systems

As anyone with even a cursory knowledge of expert systems literature would know, the dominant form of symbolic knowledge in the first generation of expert systems has been rules. In our view, the idea of rules as a formalism for encoding knowledge comes from at least three distinct traditions, and a failure to distinguish between the different senses implied by them is often a great source of misunderstanding. Let us list the three senses.

4.1. Rule Systems as "Universal" Computing Systems

It is well-known that Post productions as well as Markov Algorithms, both of which are *rule-based formalisms,* are examples of universal computation systems. Any computer program, including an expert system, can be encoded in one of these rule systems with only some minimal constraints on the interpreter. In this sense of the term, the rule-based approach becomes a *programming technology. Some* of the rules in almost every major rule-based system perform such a purely programming role. For example, rules in R1 (McDermott, 1982) which set contexts can, in another representation, be simply viewed as a call for a module which contains a block of knowledge relevant to that context. Metarules (Davis, 1976) can also be viewed as attempts to enable the specification of control behavior in a rule-based programming system. Not all algorithms can be equally naturally encoded in rule formalisms, however. Thus often expert system designers who build rule-based systems complain of frustrations they face when they have to come up with rules to make the system have good control behavior, or to express all domain knowledge in rule form; i.e., they are not encoding "domain knowledge" as much as doing *programming in a rule-system.*

Whether the rule-based approach is a good programming technology for an expert task depends on whether both the necessary control behavior as well as the "facts of the domain" can be most naturally represented in rule form. In some domains there is a 20%/80% effect — i.e., a large percentage of the domain knowledge may appear to be capturable by a relatively small number of rules, but a rapid growth in the number of rules required sets in if one attempts to capture more and more of the domain knowledge (McDermott, 1982).

4.2. Rules as in "Rules-of-Thumb"

The common sense notion of rules is one of an approximate, quick guide to action, a computationally less expensive alternative to real thinking, adequate for most occasions, but still with potential pitfalls. A related notion is that of a rule as something that captures a relationship which is only statistically valid, or a relationship whose causal antecedents are poorly understood. Rules in medical diagnosis systems of the form, "If male and findings x, y, z, assign k units of evidence to hypothesis H," are of this type, where some statistical differential between the sexes in the likelihood of having a certain disease might be used.

This sense of rules is often the basis of concerns that rule-based approaches are "shallow", and that for hard problems "causal" models are needed.

It ought to be emphasized that, this common sense notion notwithstanding, the fact that an expert system is rule-based should not necessarily imply that it is engaged in shallow reasoning, or that it is using knowledge of only approximate validity. For example, R1 uses rules which are perfectly sound pieces of knowledge about the domain of computer system configuration. To the extent that any computer program can be written in a rule-based computational framework, the shallow vs. deep characterization does not arise from its rule form, but from the character of the rules.

4.3. Rules as Cognitive Units

The third stream of ideas relating to the prominence of rules in expert systems is the idea, due to Newell (1973), of *rules* as the basic form of knowledge formation in the human short-term memory. In this theory rules become the basic building blocks of human knowledge structures. This sense of rules gives rule-based systems an aura of legitimacy as *model of intelligence*. But we feel that one should be very cautious about making this connection, even though it seems a natural one in the light of the fact that expert systems are a branch of artificial *intelligence*. This interpretation is neither necessary nor sufficient for the role of rules in expert systems. It is not necessary because, while an AI program may advantageously model human thought at some level of abstraction, it is not obvious that it should do so at the level of knowledge formation in short-term memory, especially for capturing expert problem-solving performance. It is not sufficient because even if rules were the basic units of cognition, a number of further constructs would be needed to account for their organization into higher level units such as concepts, and for their interaction with problem solving.

4.4. When Are Rule-Based Systems Appropriate?

The term "rule-based system" has come to mean an expert system which has (a) a knowledge base of rules, and (b) a problem solver — such as a forward-chaining system or a backward-chaining system, or a production system controller such as OPS-5 (Forgy & McDermott, 1977) — that uses the knowledge base to make inferences. It is conceptually important to keep in mind that the use of rules as a representation device does not necessarily force one to use this two-part rule-based system architecture, i.e., rules can be used for expert system design in significantly different architectures. MDX, for example, has some of its medical knowledge in the form of rules, but it is organized as a hierarchical collection of "specialists" (Chandrasekaran & Mittal, 1983). We shall discuss this system later in the paper.

The remarks that follow apply to the use of rules in the standard rule-based system architectures. Because the problem solver (or, the inference engine, as it has come to be called) is itself free of knowledge, controlling the problem-solving process often involves placing more or less complex rules for control purposes in the knowledge base itself. This is the "programming technology" sense of the use of rule based systems that was mentioned earlier. Normally, the rule-based architecture mentioned above works quite well when *relatively little complex coupling* between rules exists in solving problems, or the rules can be implicitly chunked into groups with little local interaction between rules in different chunks. R1 is an example where there is a virtual chunking of rules for subtasks, and the reasoning proceeds in a relatively direct and focused way. In general, however, when the global reasoning requirements of the task cannot be conceptualized as a series of linear local decisions, the rule-based systems of the simple architecture results in significant "focus" problems, i.e., since the problem solver does not have a notion of purposes at different levels of abstraction, there are often problems in maintaining coherent lines of thought. Focus needs maintenance of multiple layers of contexts, goals and plans. We shall discuss later how alternative architectures may be conceived for better focus in problem solving. These architectures will begin to erase the separation of knowledge base from the inference machinery.

There is one pragmatic aspect of rule-based approaches to expert system design that is important. Since only certain kinds of expertise have natural expression as rules, a rule-based approach may encourage a false feeling of security in capturing expertise. The human expert often may not bring up expertise hard to express in that form. The 20%/80% phenomenon is worth recalling here: i.e., as the most rule-like pieces of knowledge get encoded, there may be an acceptable initial performance, but as more knowledge that is not naturally rule-like is required, the size of the rule base may grow very rapidly.

5. On Logic and Frames as Representation Formalisms

The architecture of systems using some sort of a logical formalism for knowledge representation is generally similar to that of the rule-based systems discussed earlier. That is, there is a knowledge base of formulas in some logical formalism, and inference machinery that uses the formulas to make further inferences. But since there are several forms of logic with well-understood semantics — unlike, say, production rules — logic enjoys a status as theoretically more rigorous for knowledge representation purposes. In practical terms, however, the existence of rigorous semantics is not always helpful, since the semantics are often not at the right level of abstractions; e.g., it is often difficult to incorporate context-dependent inference strategies in logic-based systems. If one were to model, say, reasoning in arithmetic, one could represent domain knowledge in the form of axioms, and use a variety of inference machineries to derive new theorems. However, the computational complexity of such systems tends to be impractical even for relatively simple axiomatic systems. Even in such domains where in theory powerful axiom systems exist, capturing the effectiveness of human reasoning is an open research area. On the other hand, in tasks where the inference chains are relatively shallow, i.e., the discovery of the solution does not involve search in a large space, the logic representation may be more practical. Analogous to our remarks regarding the appropriateness of rule-based systems when there is an implicit structure to the task that can be mapped into chunking of rules, in logic-based systems also it is possible to create a similar subtasking structure that can keep the complexity of inference low.

To our mind, both the logic-based architectures and the rule-based architectures have surprising similarities in some dimensions; in both of them, the architecture separates a knowledge base from mechanisms that use the knowledge. In both cases, this results in an increasing need to place in the knowledge base more and more control-type knowledge — the representations increasingly become programming technologies rather than perspicuous encodings of problem solving activity. Because of an inability at that level to specify complex structures such as contexts and goal hierarchies, the approaches are subject to problems of focus in reasoning.

Control of problem solving requires, in our opinion, *organizing* knowledge into chunks, and invoking portions of the knowledge structures and operating on them in a flexible, context-dependent manner. The knowledge representation approach in AI that first emphasized organization in the form of structured units was that of *frames* (Minsky, 1975). Frames are especially useful in organizing a problem solver's "what" knowledge — knowledge about objects, in such a way that efficiency in storage as well as inference can be maintained.

Assuming some familiarity on the part of the reader with the frame concept, we will briefly mention three aspects of frames that contribute to this efficiency. Basically, a frame of a concept being a structured stereotype, much of the knowledge can be stored as "defaults," and only the information corresponding to differences from the default value needs to be explicitly stored. For example, the default value of the number of walls for a room is 4. The knowledge the system has about a particular room does not have to explicitly mention this, unless it is an exception, e.g., it is a 5-walled room.

A second benefit of organizing one's knowledge of objects in the form of frame structures is that one can create frame hierarchies, and let much of the knowledge about particular objects be inherited from information stored at the class level. This again makes for great economy of storage. For instance, the "purpose" slots of a bedroom and a living room may be different, but the parts that are common to them, e.g., typically all rooms have four walls, a ceiling etc., can be stored at the level of the "room" frame, and inherited as needed at the "bedroom" or "living room" level.

A third mechanism that makes frame structures very useful in expert systems is the possibility of embedding procedures in the frames so that certain inferences can be performed as appropriate for the conceptual context. This, as can be seen, is a move away from the rule-based system architecture, where the inference mechanisms were divorced from the knowledge base.

Because of these three properties, frame systems are very useful for capturing one broad class of problem-solving activity, viz., one where the basic task can be formulated as one of *making inferences about objects by using one's knowledge of related objects elsewhere in the structure.* Thus a whole class of programming styles called "object-oriented" programming has arisen which has conceptual kinship with the notion of frames.

6. Organization of Knowledge

Even though many knowledge-based systems follow the popular architecture of a knowledge base (generally of rules, but many times of other forms: networks in Prospector, frames in Pip) and an inference engine,[2] often they have further internal architectures. These are implicit, in the sense that they are accomplished, as mentioned in an earlier section, by using programming techniques (again, typically using rules to specify conditions for transfer to the module) to achieve some degree of modularization of the knowledge into groups. Before we discuss this, however, some account of the need for this sort of organization is called for.

When the number of domain rules in the knowledge base is large, typically several rules will "fire," i.e., their left-hand sides will match the

---

[2] We will argue in section 7 against this separation, but for our immediate purposes that is not important.

state of the data. The inference machinery (because in the standard architecture it is deliberately kept domain-independent) does not have the domain knowledge to choose among them. Therefore, either some sort of syntactic conflict-resolution mechanisms need to be used (such as the technique of R1, which chooses that rule whose conditions strictly subsume those of a contending rule), or all of the rules will need to be tried. The latter option has the potential for combinatorial explosion. Most systems attempt to cope with this problem by creating "contexts," which help specify a small set of rules in the base as candidates to be considered for matching. For example, each subtask in R1 creates a context and only rules relevant to that subtask are considered for matching. This technique essentially decomposes the rule base (except for the rules which effectuate the transfer from module to module) into a number of virtual modules, each for a subtask. PROSPECTOR (which is a geological consultation system, [Duda et al., 1979]), while not a rule-based system explicitly,[3] also organizes its knowledge in the form of "models," each model corresponding to a classification hypothesis about the geological make-up. In a sense, each model is a "specialist" in that hypothesis. Metarules have been proposed as a special class of rules to embody "control knowledge" (Davis, 1976). These also play the role of decomposing the knowledge base into portions that are relevant for classes of situations. Without some such attempt at organization, the problem-solving process will generally be very *unfocused,* and serious control problems will arise.

All the above organizational devices are *implicit,* and are subject to the constraints of the rule formalism on the one hand, and the uniformity of the inference procedure on the other. The uniformity of the inference machinery makes it difficult for different subtasks during reasoning to exploit different ways of going about using the knowledge. Again it is worth emphasizing that the issue is not one of the computational sufficiency of the rule mechanisms, but one of naturalness and conceptual adequacy.

Our own work (e.g., (Chandrasekaran, 1983), which gives an overview of our activity) has been directed toward the development of a *theoretical basis* of *knowledge organization* for expert problem solving. We will outline some aspects of this theory in the next section.

7. Concepts and Problem Solving Types as Organizational Constructs

To restate a point from the last section: even the implicit modularization of the knowledge base due to various context-setting mechanisms is not

---

[3] The *content* of the network representation can be translated into rule forms in a straightforward way.

always sufficient when the task domain consists of subtasks which might differ from each other in the *nature* of the problem solving, i.e., the *use* of knowledge that is required for that subtask. The work to be discussed has been directed toward elaborating a framework in which different generic types of problem solving can be related to the types of knowledge organizations required for them.

7.1. Generic Tasks

The theory proposes that there are well-defined *generic tasks* each of which calls for a certain organizational and problem-solving structure. We have identified several such generic tasks from our work in the domains of medicine and reasoning about engineered systems.

The classificatory task that is at the core of medical diagnosis, i.e., the task which classifies a complex case description as a node in the disease hierarchy, is an example of such a generic task. It is generic because it is a component of many real world problem solving situations. For example, a tax-advising expert program might go through a stage of classifying the user as a particular *type* of taxpayer before invoking strategies that may be appropriate for that class of taxpayer. We have already discussed in section 3 simpler versions of this task. The problem solving for this task, as implemented in our medical diagnosis system, MDX, will be considered in section 7. 1. 1.

Another example of a generic task uses what we call WWHI-type (for "What Will Happen If") reasoning, which attempts to derive the consequences of an action that might be taken on a complex system. Such a task is useful as a subtask in an expert system that trouble-shoots and repairs a complex system, where it may be useful to reason out the consequences of a proposed corrective action.

A third type is a form of knowledge-directed associative memory that helps retrieve information by reasoning about other related information; we have used this type of problem solving in an intelligent database system, PATREC (Mittal & Chandrasekaran, 1980). A fourth type is a form of plan synthesis, which we are using to build an expert system for mechanical design (Brown & Chandrasekaran, 1983). It is clear that there are many more such generic tasks, and it is part of our research program to identify more of them.

An important consequence of identifying such tasks is that it gives us a framework to characterize the capabilities of expert systems. If a real world task can be decomposed into a number of generic tasks, and if for each of them we know how to build a reasoning system, then there will be a basis for concluding that the task domain can be successfully tackled by an expert system.

In the next two subsections, we will discuss in greater detail, but still in schematic terms, how knowledge can be organized and problem solving can be accomplished for two of the above generic tasks. Cited references can be consulted both for more details on these tasks, as well as for information on the other two problem-solving types that we omit here owing to space limitations.

7.1.1. The Classificatory Task. As mentioned earlier, the task is the identification of a case description with a specific node in a predetermined diagnostic hierarchy. For the purpose of current discussion let us assume that all the data that can be obtained are already there, i.e., the additional problem of launching exploratory procedures such as ordering new tests etc. does not exist. The following brief account is a summary of the more detailed account given in Gomez and Chandrasekaran (1981) of diagnostic problem solving.

Let us imagine that corresponding to each node of the classification hierarchy alluded to earlier we identify a "concept." The total diagnostic knowledge is then distributed through the conceptual nodes of the hierarchy in a specific manner to be discussed shortly. The problem solving for this task will be performed top down, i.e., the top-most concept will first get control of the case, then control will pass to an appropriate successor concept, and so on. In the medical example, a fragment of such a hierarchy might be as shown in Figure 1.

More general classificatory concepts are higher in the structure, while more particular ones are lower in the hierarchy. It is as if INTERNIST first establishes that there is in fact a disease, then LIVER establishes that the case at hand is a liver disease, while say HEART etc. reject the case as being not in their domain. After this level, JAUNDICE may establish itself and so on.

The problem solving that goes on in such a structure is *distributed.* The problem-solving regime that is implicit in the structure can be characterized as an *"establish-refine"* type. That is, each concept first tries to

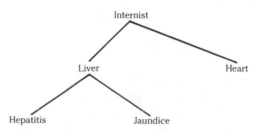

Figure 1. Fragment of a classificatory hierarchy.

establish or reject itself. If it succeeds in establishing itself, the refinement process consists of seeing which of *its* successors can establish itself. The way in which each concept (or "specialist") attempts to do the establish-refine reasoning may vary from domain to domain. In medicine it may often be accomplished by using knowledge in the form of a collection of rules, some of which look for evidence for the hypothesis, some for counter evidence, and others which carry information about how to combine them for a final conclusion. In reasoning about electrical circuits on the other hand it may be more appropriate to represent the establish-refine activity in the form of functional knowledge about specific modules. (That is, performance of a generic task may require solution of some problem of a different type as a subtask.)

In our medical diagnosis system MDX, each of the concepts in the classification hierarchy has "how-to" knowledge in it in the form of a collection of *diagnostic rules.* These rules are of the form: <symptoms> ----->
<concept in hierarchy>, e.g., "If high SGOT, add n units of evidence in favor of cholestasis." Because of the fact that when a concept rules itself out from relevance to a case, all its successors also get ruled out, large portions of the diagnostic knowledge structure never get exercised. On the other hand, when a concept is properly invoked, a small, highly relevant set of rules comes into play.

Each concept, as mentioned, has several clusters of rules: confirmatory rules, exclusionary rules, and perhaps some recommendation rules. The evidence for confirmation and exclusion can be suitably weighted and combined to arrive at a conclusion to establish, reject or suspend it. The last mentioned situation may arise if there is not sufficient data to make a decision. Recommendation rules are further optimization devices to reduce the work of the subconcepts. Further discussion of this type of rule is not necessary for our current purpose.

The concepts in the hierarchy are clearly not a static collection of knowledge. They are active in problem solving. They also have knowledge only about establishing or rejecting the relevance of that conceptual entity. Thus they may be termed "specialists," in particular, "diagnostic specialists." The entire collection of specialists engages in distributed problem solving.

The above account of diagnostic problem solving is quite incomplete. We have not indicated how multiple diseases can be handled within the framework above, in particular when a patient has a disease secondary to another disease. Gomez has developed a theory of diagnostic problem solving which enables the specialists in the diagnostic hierarchy to communicate the results of their analysis to each other by means of a *black-board.* Thus the problem solving by different specialists can be coordinated. Similarly, how the specialists combine the uncertainties of medical data and diagnostic knowledge to arrive at a relatively robust conclusion about establishing or rejecting a concept is an important issue, for a discussion of which we refer the reader to Chandrasekaran, Mittal and Smith (1982).

The points to notice here are the following. The inference engine is tuned for the classificatory task, and the control transfer from specialist to specialist is implicit in the hierarchial conceptual structure itself. One could view the inference machinery as "embedded" in each of the concepts directly, thus giving the sense that the concepts are "specialists."

7.1.2. *What-Will-Happen-If* (WWHI) *or Consequence* Finding. Examples

of this type of reasoning are: "What will happen if valve A is closed in this power plant when the boiler is under high pressure?"; "What will happen if drug A is administered when both hepatitis and arthritis are known to be present?" Questions such as these can be surprisingly complex to answer since formally they involve tracing a path in a potentially large state space. Of course what makes it possible in practice to trace this path is domain knowledge which constrains the possibilities in an efficient way.

The problem solving involved, and correspondingly the use of knowledge in this process, are different from that of diagnosis. For one thing, many of the pieces of knowledge for the two tasks are completely different. For example, consider answering the question in the automobile mechanic's domain: "What will happen if the engine gets hot?" Looking at all the diagnostic rules of the form, "<hot engine ----> <malfunction>" will not be adequate, since <malfunction> in the above rules is the *cause* of the hot engine, while the consequence finding process looks for the effects of the hot engine. Formally, if we regard the underlying knowledge as a network connected by cause-effect links, where from each node multiple cause links as well as effect links emanate, we see that the search processes are different in the two instances of diagnosis and consequence-finding. The diagnostic concepts that typically help to provide *focus* and constrain search in the pursuit of correct causes will thus be different from the WWHI concepts needed for the pursuit of correct effects.

The embedded problem solving is also correspondingly different. We propose that the appropriate language in which to express the consequence-finding rules is in terms of *state-changes*. To elaborate:

1. WWHI-condition is first understood as a state change in a subsystem.
2. Rules are available which have the form "<state change in sub-system> will result in <state change in subsystem>". Just as in the case of the diagnosis problem, there are thousands of rules in the case of any nontrivial domain. Again, following the diagnostic paradigm we have already set, we propose that these rules be associated with *conceptual specialists*. Thus typically all the state change rules whose left hand side deals with a subsystem will be

aggregated in the specialist for that subsystem, and the right-hand side of those rules will refer to the state changes of the *immediately affected* systems.

Again we propose that typically the specialists be organized hierarchically, so that a subsystem specialist, given a state change, determines by knowledge-based reasoning the state changes of the immediately larger system of which it is a part and calls that specialist with the information determined by it. This process will be repeated until the state change(s) for the overall system, i.e., at the most general relevant level of abstraction, are determined. This form of organization of the rules should provide a great deal of focus to the reasoning process.

An Illustrative Example. Consider the question, in the domain of automobile mechanics, "WWHI there is a leak in the radiator when the engine is running?" We suggest the specialists are to be organized as in Figure 2:

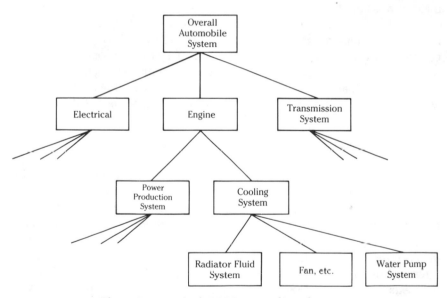

Figure 2. *Example of WWHI concept hierarchy.*

The internal states that the *radiator fluid subsystem* might recognize may be partially listed as follows: {leaks/no leaks, rust build-up, total amount of water, ... }; similarly the *fan subsystem* specialist might recognize states {bent/straight fan blades, loose/tight/disconnected fan belt, ... }. The *cooling system subsystem* itself need not recognize states to this degree of detail; being a specialist at a somewhat higher level of abstraction it will recognize states such as {fluid flow rate, cooling-air flow rate ... etc.}. Let us say that the *radiator fluid* specialist has, among others, the following rules. The rules are typically of the form: <internal state change> ----> <supersystem state change>.

> leak in the radiator ----> reduced fluid flow-rate
> high rust in the pipes ----> reduced fluid flow-rate
> no antifreeze in the water and very cold weather ----> zero fluid flow etc.

The cooling system specialist might have rules of the form:

> low fluid-flow rate and engine running ----> engine state hot
> low air-flow rate and engine running ----> engine hot

Again note that the internal state recognition is at the appropriate level of abstraction, and the conclusions refer to state changes of its parent system.

It should be fairly clear how such a system might be able to respond to the query about radiator leak. Again, a blackboard for this task would make it possible to take into account subsystem interaction.

Unlike the structures for the diagnostic and data retrieval tasks, we have not yet implemented a system performing the WWHI-task. While we cannot speak with assurance about the adequacy of the proposed solution, we feel that it is of a piece with the other systems. That is, it illustrates embedding still another type of problem solving in a knowledge structure which consists of cooperating specialists of the same problem-solving type.

7.2. Discussion

Since each of the generic tasks involves a problem-solving behavior which is unique to that task, the standard architecture of a knowledge base and a general purpose inference machinery is not applicable here. There is a closer intertwining of knowledge structures and corresponding inference methods. At the implementation level, one can view the system as being decomposed into a collection of pairs of the form (<knowledge structure, inference method>), indexed by the generic tasks, e.g., <diagnostic structure, establish-refine>. However it is conceptually more appropriate to view each of the specialists as having the inference machinery "embedded" in them. This interpretation gives the term "specialists" an added degree of aptness.

In section 2 we mentioned that the first generation of expert systems emphasized the power of knowledge itself over that of the problem-solving method. In the current section we have attempted to restore the balance, by showing how a variety of problem-solving types in conjunction with appropriate organizations of knowledge can solve a greater variety of problems.

8. Expert Systems that Understand

We have outlined an evolution of expert systems from collections of rules to cooperative problem solving by a community consisting of specialists in different kinds of problem solving. The knowledge that is in each of the specialists, e.g., the diagnostic rules in the classificatory specialists, or the state abstraction rules in the WWHI specialists in the previous section, is itself "compiled." This knowledge is obtained from human experts who either learnt that knowledge originally in that compiled form, formed it as a result of experience, or *derived* it from a deeper model of the domain. In Chandrasekaran and Mittal (1982) we argue that in principle, given any deep model of the domain, one can compile an MDX–like diagnostic system which is as powerful as the deeper model, but more efficient than it, for the diagnostic problem. However, in practice, the compiled structures are likely to remain *incomplete* for various reasons, and it would be very attractive to endow expert systems with deeper understanding of their domains to protect themselves against incompleteness.

Attempts to give expert systems some ability to do deeper level reasoning have typically taken the direction of giving the system a mechanism to reason at different levels of detail by using a prestored knowledge base of causal associations. In CASNET (Weiss, et al., 1978) an attempt is made to trace out the most likely causal sequences given some likely intermediate states for which there is evidence, and some states which can be assumed not to have occurred. A knowledge base of possible causal connections between states with associated likelihood information is used to fit a most likely path that goes through the states for which there is evidence and avoids the unlikely states. ABEL (Patil, 1981) uses causal association information at different levels of detail. There may be a piece of knowledge at the top level of the form, "A causes B," but at a more detailed level, there may be several different ways in which A might be able to cause B. The system works at these different levels of detail to pin the causal connection down to the degree that is required.

It is important to note that these systems do not use "causal models" as much as they use a storehouse of compiled causal associations. In a sense all diagnostic programs use "causal models." Much of the diagnostic knowledge in MYCIN or MDX *is* causal, i.e., saying "symptom A gives so much evidence for disease B," is, in content, the same as "B causes A with so much likelihood." The difference between them and the programs above is what is done with the causal knowledge.

In our view a truly deep model should have the power to *derive* the causal connections between states. The work of Kuipers (1982), deKleer and Brown (1981) and Moorthy and Chandrasekaran (1983) are relevant here. Kuipers proposes methods by which causal behavior may be derived from a knowledge of the structure of some system. The latter two references seek to model functional understanding of devices.

The scope of the current paper does not allow a detailed description of how the functional representations work and how they are related to diagnostic reasoning. Here we content ourselves with an intuitive account of our approach.

We model *understanding of a device* as the creation of a knowledge structure which is hierarchically organized in terms of functions and subfunctions. How the *salient behavior* (generally stated in qualitative terms) and *the physical structure* of the relevant portions of the device play a role in achieving a function by means of the subfunctions is part of such a description.

We believe that such a structure can be used to generate the causal knowledge needed for diagnostic reasoning. The function/subfunction relationship can be used to generate diagnostic hypotheses. If function A is affected, each of its subfunctions can be considered as a possible source of failure. Similarly, the symptomatic knowledge that is needed for establishing or rejecting these as possibilities can be derived from the behavioral and physical structure constraints that enable subfunctions to be achieved. The attempt to give systems a degree of understanding based on functional models is still very experimental, and practical expert systems based on this approach are still some time away. But we feel that this sort of systems is the next step in the evolution of expert systems towards a greater degree of understanding.

9. Discussion

The reader may have gathered that there is no simple method of determining which tasks are likely to be successfully handled by an expert system. We have attempted to give some analytical tools by which such a decision may be made a little easier. In any case, the following guidelines arise from our experience in designing a number of expert systems.

1. When the total amount of knowledge is relatively small (a few hundred rules), the exact technique used is not very important. A wide variety of techniques will all give similar qualitative performance.
2. A large fraction of those expert systems that have reached some degree of exposure (PROSPECTOR, MYCIN, MDX, CASNET, INTERNIST) deal with some form of the classification problem. If the core task in an application domain is classificatory, chances are very good that an expert system approach will be successful. Problems

of synthesis, such as design, are in general harder, but some simpler versions of the design problem, such as the task domain of R1, have been successfully attacked.

3. Another type of expert system that is likely to be of practical applicability is one that helps the user access knowledge in a complex knowledge base. This type of expert behavior does not require the full problem solving capabilities of the expert. Our work on intelligent databases (Mittal and Chandrasekaran, 1980) offers some techniques of potential relevance here.

4. If a real-world task can be decomposed into a number of generic tasks for which expert system solutions are available, then the prospect of a successful expert system increases significantly.

5. Important research is going on in common sense qualitative reasoning involving space and time; these advances will give the next generation of expert systems more power and flexibility.

6. Reasoning of experts is in general varied and broad-ranging. We have only begun to understand some forms of such reasoning. Honesty compels us to admit that it is not a simple matter to capture all forms of expertise and incorporate them in the form of computer programs, even though, in the enthusiasm surrounding this promising field, careful distinctions and qualifications often do not get made. On the positive side is the solid body of accomplishment: the field *has* managed to capture a number of useful forms of expertise.

We have not discussed in this paper a number of issues such as user interfaces, explanation facilities, and knowledge acquisition problems. They are obviously of great practical importance. But we believe the issues of knowledge organization and problem solving will continue to occupy the center stage in this area, since these problems are by no means solved.

10. References

Brown, D. C., & Chandrasekaran, B. Expert systems for mechanical design. *Proceedings IEEE Trends & Applications in Artificial Intelligence,* 1983, 173–180.

Chandrasekaran, B. Toward a taxonomy of problem solving. *AI Magazine,* Winter/Spring 1983, *IV*(1), 9–17.

Chandrasekaran, B., & Mittal, S. Deep versus compiled knowledge approaches to diagnostic problem-solving. *Proceedings of the American Association of Artificial Intelligence Conference,* 1982, 349–354. To appear in *International Journal of Man Machine Studies,* 1983.

Chandrasekaran, B., & Mittal, S. Conceptual representation of medical knowledge. In M. Yovits (Ed.), *Advances in Computers,* (Vol. 22). New York: Academic Press, 1983.

Chandrasekaran, B., Mittal, S., & Smith, J. W. M. D. Reasoning with uncertain knowledge: the MDX approach. *Proceedings of the First Annual Joint Conference of the American Medical Informatics Association,* 1982, 335–339.

Davis, R. *Applications of metalevel knowledge to the construction, maintenance, and use of large knowledge bases.* Unpublished doctoral dissertation, Stanford University, 1976.

Davis, R. Expert systems: Where are we? And where do we go from here? *AI Magazine,* Spring 1982, *3*(2), 3–22.

deKleer, J., & Brown, J. S. Mental models of physical mechanisms and their acquisition, In Anderson (Ed.), *Cognitive Skills and Their Acquisition.* Hillsdale, NJ: Lawrence Erlbaum Associates, 1981.

Duda, R. O., Gaschnig, J., & Hart, P. E. Model design in the prospector consultant system for mineral exploration. In D. Michie (Ed.), *Expert Systems in the Micro-Electronic Age.* Edinburgh University Press, 1979.

Forgy, C., & McDermott, J. OPS: A domain-independent production system. *Proceedings of the Fifth International Joint Conference on Artificial Intelligence,* 1977, 933–939.

Gomez, F., & Chandrasekaran, B. Knowledge organization and distribution for medical diagnosis. *IEEE Trans. Systems, Man, and Cybernetics,* Vol. SMC-11, No. 1, January 1981, 34–42.

Kuipers, B. *Commonsense reasoning about causality: Deriving behavior from structure* (Working Papers in Cognitive Science, No. 18). Tufts University, 1982.

Marr, D. Early processing of visual information. *Phil. Trans. of the Royal Society of London,* Vol. 275, Series B, 1976, 483–524.

McDermott, J. RI: A rule-based configurer of computer systems. *Artificial Intelligence,* 1982, *19* (1), 39–88.

Minsky, M. A framework for representing knowledge. In P. H. Winston (Ed.), *The Psychology of Computer Vision.* New York: McGraw Hill, 1975.

Minsky, M., & Papert, S. *Perceptrons: An introduction to computational geometry.* Cambridge, MA: MIT Press, 1969.

Mittal, S., & Chandrasekaran, B. Conceptual representation of patient data bases. *Journal of Medical Systems,* 1980, *4* (2), 169–185.

Moorthy, V. S., & Chandrasekaran, B. A representation for the functioning of devices that supports compilation of expert problem solving structures. *Proceedings of the IEEE MEDCOMP 83,* IEEE Computer Society 1983.

Newell, A. Production systems: Models of control structure. In W. Chase (Ed.), *Visual Information Processing.* New York: Academic Press, 1973.

Newell, A., & Simon, H. A. *Human problem solving.* Englewood Cliffs, NJ: Prentice-Hall, 1972.

Nilsson, N. J. *Problem solving methods in artificial* intelligence. New York: McGraw-Hill, 1971.

Patil, R. S. *Casual representation of patient illness for electrolyte and acid-base diagnosis.* Unpublished doctoral dissertation, TR–267, Cambridge, MA: MIT Lab for Computer Science, October, 1981.

Pauker, S. G., Gorry, G. A., Kassirer, J. P., & Schwartz, W. B. Towards the simulation of clinical cognition. *The American Journal of Medicine,* 1976, *60,* 981–995.

Samuel, A. Some studies in machine learning using the game of Checkers. *IBM Journal of Research & Development,* 1967, 601–617.

Stefik, M., Aikins, J., Balzer, R., Benoit, J., Birnbaum, L., Hayes-Roth, F., & Sacerdoti, E. *The organization of expert systems: A prescriptive tutorial* (Tech. Rep.). Xerox PARC, 1982.

Szolovits, P., & Pauker, S. G. Categorical and probabilistic reasoning in medical diagnosis. *Artificial Intelligence,* 1978, *11,* 115–144.

Weiss, S. M., Kulikowski, C. A., Amarel, S., & Safir, A. A model-based method for computer-aided medical decision-making. *Artificial Intelligence,* 1978, *11,* 145–172.

Coupling Expert Systems With Database Management Systems*

Matthias Jarke and Yannis Vassiliou
GRADUATE SCHOOL OF BUSINESS
ADMINISTRATION
NEW YORK UNIVERSITY

The combined use of database management systems (DBMS) and artificial intelligence-based Expert Systems (ES) is potentially very valuable for modern business applications. The large body of facts usually required in business information systems can be made available to an ES through an existing commercial DBMS. Furthermore, the DBMS itself can be used more intelligently and operated more efficiently if enhanced with ES features. However, the implementation of DBMS–ES cooperation is not easy.

We explore practical benefits of the cooperative use of DBMS and ES, as well as the research challenges it presents. Strategies for providing data from a DBMS to an ES are given. Complementary strategies for providing intelligence from an ES to a DBMS are also presented. Finally, we discuss architectural issues such as degree of coupling, and combination with quantitative methods.

As an illustration, a research effort at New York University to integrate a logic-based business ES with a relational DBMS is described.

1. Introduction

In the mid-1960s the world watched with fascination a volcano off the fishing town of Heimaey, Iceland, create a new island, Surtsey, out of its ashes. Such things had happened before, and this particular creation process went

* The work reported in this chapter is supported in part by a joint study with International Business Machines Corporation. Jim Clifford designed and implemented the internal relational database system in PROLOG and codeveloped the staged approach to getting data into the expert system. Taracad Sivasankaran did a major part of the initial design and preliminary implementation of the insurance expert system used as an example in this paper. We also thank the other project members, Hank Lucas, Ted Stohr, Norm White, and especially Walt Reitman for encouragement and many helpful discussions. Finally, we thank Sibylle Hentsch of Hamburg University for acquainting us with the intricacies of volcanoes off Iceland.

on for four years. But, unlike on previous occasions, the new island did not vanish in the sea shortly after its creation but seemed determined to stay. Still, it is a rather tiny little island made visible to the world mainly by the big volcanic clouds that accompanied its creation.

Some people began to populate the new island and the problem arose how to provide them with the necessary things of life from the neighboring islands (Iceland being the most important). On the other hand, researchers from all over the world flocked in to unravel the hidden treasures of this new piece of land in the hope to clarify and resolve some of the problems in other areas. Currently, fishing boats manage the traffic but soon something more solid will be required to cope with the growing interaction.

What does all this have to do with expert systems and databases? The reader might have noticed a strong resemblance of Surtsey's development with the emergence of expert systems over the past few years. Once a small offshot of artificial intelligence, they now catch the spotlight of publicity after a prolonged period of laboratory existence.

Much research has been necessary to improve expert systems to commercial feasibility. With the actual intrusion of expert systems into the business world, however, the further question arises how to integrate them into the existing Management Information Systems (MIS).

Can bridges (or at least fishing boats) be built as a connection to the other subsystems that supply the expert systems with the necessary business data and allow them to use corporate planning models and other mathematical models as a human expert would?

Can, vice versa, the expert system idea be used to enhance existing MIS with more intelligence in systems design, usage, and operation?

Researchers who have addressed these problems typically looked at certain aspects of one of the two questions. In this paper, we try to view them together, with a primary focus on the interaction of expert systems with large existing databases. To remain in our original picture, we try to outline a bridge between the two systems that allows for two-way traffic enhancing the quality and efficiency of both types of systems.

The paper will analyze on a high level opportunities and solution strategies for getting business data from a commercial database system into an expert system, and for providing expert knowledge to database systems. For more technical details of the first problem, the reader is referred to a companion paper (Vassiliou et al., 1983) and to some work in the Japanese Fifth Generation Computer project (Kunifuji & Yokota, 1982).

Certain aspects of the second problem are addressed in (Jarke & Koch, 1983), (Jarke & Shalev, 1984), as well as in the work of many others (such as Reiter, 1978; Chang, 1978; Kellogg, 1982; King, 1979; and Henschen et al., 1982). However, the contribution of this paper is to unify all these approaches in a common architectural framework of what we believe might be the paradigm of forthcoming advanced decision support systems.

For an illustration, we will use from time to time examples from an ongoing project at New York University in which a logic-based business expert system is developed and interfaced with an existing relational database and mathematical subroutine libraries.

The paper is organized as follows. Section 2 gives a brief review of major subsystems and requirements in a business. In subsequent sections, we study the interaction between these components, focusing on the relationships between databases and expert system knowledge bases and inference engines. Section 3 presents a stage-wise approach to the question of how to get data for a business expert system. Section 4 deals with various applications of expert system technology to the design, use, and operation of database systems. Finally, section 5 addresses some architectural and technical problems of the coupling between expert and database management systems that have to be solved regardless of the direction of the interaction.

2. Review of Business Subsystems and Requirements

In this section, the "islands" to be connected in a business will be identified, and their interaction requirements investigated. One type of current business systems is characterized by backbone transaction processing, governed by rules of what can be called a business program, and typically centered around large centralized or distributed databases. The knowledge required for building and working with such systems is typically hidden in procedural form and cannot be easily changed nor carried over to other similar applications. Modern transaction processing system types, such as office automation systems, make these rules more explicit and formalize the notion of documents as a central carrier of information to be handled (Tsichritzis, 1982).

Another type of business systems deals more directly with supporting decisions on various levels. Decision support systems often have their knowledge built in mathematical formulas and models which are handled by model management algorithms, or by the user via a flexible and powerful user interface (Stohr & White, 1982).

Finally, there is a large number of human specialists who use experience and expertise together with factual knowledge gained from databases via query languages to develop recommendations and explanations in their area of expertise. Recently, expert systems have been devised that efficiently support or partially replace such specialists. Expert systems typically organize their knowledge in the form of IF-THEN rules and perform some form of pattern matching to find out which rules apply.

To summarize this discussion, it can be said that businesses use: (a) large databases, (b) mathematical formulas and models, (c) business rules and forms (documents), and (d) experience and expertise of various human specialists. If expert systems are introduced, there will be another type of subsystem which interacts with the human decision-makers to complement their expertise.

In this kind of setting it is not hard to see that much effort is duplicated if all these systems operate independently of each other.

Today, most mathematical and expert subsystems have their own data management facilities. If data from a database system are used they are typically extracted as snapshots before the beginning of a session. This approach causes problems if the amount of data to be extracted is large or cannot easily be determined in advance. If the snapshots are kept over an extended period of time, there is a problem of keeping them consistent with the main database.

Database management systems usually offer some mathematical (e.g., aggregation) capabilities in their user interfaces and elementary rule-based techniques (integrity constraints) to check their operations. However, often the user would like more sophisticated retrieval facilities which perform reasoning or mathematical operations on the data before presenting them to the user. A higher level of semantic knowledge and deductive capabilities built into the database system would not only make it more user-friendly but also safer and more efficient to operate.

Finally, on a higher level, it can be observed that system development effort is duplicated on a large scale in the analysis and design of business information systems. This happens within but even more among organizations, when systems of the same type are developed again and again without taking explicit advantage of knowledge from previous experiences.

Some research has been done to integrate the separate subsystems. Decision support systems have to base their mathematical modelling capabilities on solid database management technology to become cost-effective. By analogy, it can be argued that many expert systems which use a large population of specific facts need a communication channel to the corporate databases. Going a step further, we argue that future decision support systems will have to integrate all three components: database, mathematical subsystem, and knowledge base.

As an example for such a system, consider a life insurance consulting expert system currently under development in our group. The system develops, recommends, and explains customized life insurance policies for a customer. The system will assist a sales agent by

1. extracting customer information from the corporate database and from interviews to deduce the needs for life insurance coverage in different stages of the life of a customer;
2. classifying the requirements in terms of basic actuarial types of insurance, and relating them to applicable actuarial models in the mathematical subsystem;
3. computing a premium for the customized policy, reducing it by already existing coverage from previous insurance, and comparing the remainder to the premium paying capabilities of the customer;
4. analyzing the feasibility of the proposed policy in terms of legal requirements and corporate objectives; and
5. interacting with the customer to come up with an acceptable policy that satisfies the customer's needs and has an affordable premium.

Currently, customized policies are feasible only for major group policies since there is a lack of actuarial experts to support a process as outlined above. Individual customers can essentially only choose among a limited number of prepackaged policy combinations (insurance products). When operational, the expert system will thus relieve a bottleneck which seriously impedes service quality.

On the other hand, such a system needs:

1. information about customers and actuarial table data from a database system;
2. a mathematical subsystem to execute actuarial computations efficiently;
3. an (extensible) knowledge-based expert subsystem to extract information, to classify insurance needs, to check time-varying legal and corporate constraints, and to explain problems and alternatives.

We are implementing such a system using the logic language PROLOG, a relational database system, and a mathematical subroutine library. This project is meant to serve as a demonstration of a typical architecture we expect for upcoming knowledge-based business decision support systems.

In the remainder of this paper, we shall focus on the interaction of two of the subsystems discussed in this section, namely, knowledge-based expert systems and databases. First, we explore alternatives for extracting data from large and/or existing databases—getting supply to the inhabitants of the new island, expert systems. Later on, we turn our attention the other way: how can existing subsystems, especially database management systems, be improved when expert system technology becomes available?

3. Data for Expert Systems

In a rule-based expert system, the "inference engine" uses a set of rules (the *knowledge base*) and a collection of specific facts *(database)* to simulate the behavior of a human expert in a specialized problem domain.

For instance, the life insurance consulting expert system of section 2 uses rules like:

> IF C is a customer of a certain age and when the customer dies a certain amount becomes payable, THEN the customer is said to have a whole life insurance benefit.

It also uses a database containing data about actual customers, insurance and annuity benefits, mortality values, etc.

A database is represented in terms of two basic dimensions: variety and population. For example, in a logic-based representation the different logic predicates ("customer", "covered by", etc.) reflect the *variety,* and the instances of these predicates ("Smith is a customer", "Smith is covered by term annuity", etc.) refer to the *population.*

Most expert system databases exhibit a large variety of facts. In contrast, the population of facts in such databases is more variable; ranging from a small "laboratory" set to a very large collection of facts. Thus, expert system databases differ from traditional commercial databases in that they tend to be more "wide" and less "deep".

In Vassiliou et al. (1983), we examine the problem of expert system database representation and retrieval. We now present a summary of our results. We start with an outline of a four-stage approach in section 3.1. Then, for an illustration, we briefly describe in section 3.2. a working implementation of this approach in the programming language PROLOG.

3.1. Four-Stages of Database Management Development

Four strategies for establishing a cooperative communication between the deductive and data components of an expert system (ES) have been identified. Starting from elementary facilities for data retrieval we progress to a generalized database management system (DBMS) within the expert system, to a 'loose' coupling of the ES with an existing commercial DBMS, and finally, to a 'tight' coupling with an external DBMS. Expert system designers may opt for one configuration over another depending on data volume, multiplicity of data use, data volatility characteristics (how frequently data change), or data protection and security requirements. Regardless, in a careful design these enhancements are incremental, allowing for a smooth transition from a less to a more sophisticated environment.

3.1.1. Stage 1: Elementary Data Management within the ES. In the simplest case, all data is kept in core and stored in mostly ad-hoc data structures. Application-specific routines for data retrieval and modification are implemented.

3.1.2. Stage 2: Generalized Data Management within the ES. When the expert system database is large enough not to fit in core, elementary data management is not sufficient. Techniques are needed for external file management (e.g., secondary indexes, data directory, etc.). These techniques should preferably be application independent, since otherwise a small change in application descriptions may require an altogether different mechanism.

Furthermore, depending on the multiplicity of database use and the extent of fact variety required for the ES, general purpose database management facilities may be needed. Such facilities include "views", or dynamic database windows, available in most modern relational DBMS.

Another facility is an integrated data dictionary that allows for queries about the database structure. In a nutshell, a generalized DBMS integrated in the expert system may be necessary to deal effectively with large databases.

This stage of enhancing database management facilities seems to be the norm for expert systems that deal with large databases, even though not all such systems exhibit the same level of sophistication.

3.1.3. Stage 3: Loose Coupling of the ES with an Existing DBMS. Often, the need for consulting *existing* very large databases arises. Such databases will normally be managed by a commercial DBMS. In a typical example, Olson and Ellis (1982) report experience with PROBWELL, an expert system used to determine problems with oil wells. A very important source of information for such determination is a large database stored under the IMS database system. Unfortunately, this database cannot be made available to the ES in a timely manner.

Existing external databases are typically very large, highly volatile, and used by several applications. Costs of storing data and maintaining consistency may prohibit the duplication of such a database for the sake of the expert system alone.

Loose coupling of an ES with an existing DBMS refers to the presence of a communication channel between the two systems which allows for data extraction from the existing database, and subsequent storage of this "snapshot" as an expert system database. Data extractions occur *statically* before the actual operation of the ES.

We note that the availability of generalized database management facilities within the expert system may greatly facilitate this process. After the internal expert system database has been expanded with an external database snapshot, it can be accessed as in stage 2.

3.1.4. Stage 4: Tight Coupling of an ES With an Existing DBMS. The main disadvantage of loose coupling is the nonapplicability in cases where automated *dynamic* decisions, as to which database portion is required, are necessary. During the same ES session, many different portions of the external database may be required at different times; the requirements may not be predictable. In addition, if the external database is continuously updated, for instance in databases for commodity trading or ticket reservation, the snapshots become rapidly obsolete.

Tight coupling of an ES with an external DBMS refers to the use of the communication channel between the two systems in such a way that the external database appears to the ES as an extension of its own. Clearly, the most important consideration for the implementation of tight coupling is the "intelligent" management of the communication channel—*when* and *how* to use the channel.

Our suggested strategy, which we describe in more detail in section 3.2.4., assumes the existence of a high-level, sophisticated mechanism within the ES—an expert system itself—that collects ES requests for data while simulating the ES deduction process. This expert extracts and translates the necessary data. The facilities described in stages 2 and 3 are necessary for the implementation of tight coupling.

This completes our general discussion of getting data into expert systems. In the following subsection, we describe an application of this stage-wise approach to the problem of data storage and retrieval for a PROLOG-based expert system using relational database management systems. Most of the mechanisms described below have been implemented and tested in experimental PROLOG programs presented in Vassiliou et al. (1983).

3.2. Illustration—Data For A Prolog-based Expert System

PROLOG is a programming language based on a subset of first-order logic, the Horn clauses. Roughly, this amounts to dropping disjunction from logical consequents, and talking only about definite antecedent-consequent relationships. There are three basic statements in PROLOG (Nau, 1983):

:− P.	means	P is a goal to be proved
P.	means	P is an assertion
P :− Q,R,S	means	Q and R and S imply P

A PROLOG program is a sequence of clauses whose variables are considered to be universally quantified. A clause has both a declarative and a procedural interpretation. Thus,

P :− Q.R.S.

can be read declaratively:

P is true *if* Q and R and S are true

or, procedurally:

To satisfy P first satisfy Q and R and S.

Since more than one clause may be needed to define a predicate (goal), there is a corresponding AND/OR graph for each predicate. The execution of a program involves a depth-first search with backtracking on these graphs, and uses the unification process based on the resolution principle (Robinson, 1965).

A knowledge base can be represented in first-order logic if the formulas are suitably interpreted. Therefore, PROLOG may be used for the knowledge representation. Furthermore, PROLOG has the advantage that it already has a very powerful inference engine in place. The unification algorithm used in PROLOG is more powerful than simple pattern matching algorithms common in production systems.

3.2.1. Elementary Data Management—Naive Use of PROLOG. On the simplest level of database management facilities enhancements, stage 1, the whole population of facts can be represented directly as the expert system database. While this approach is feasible for any expert system, using PROLOG can take us a step further. Since PROLOG does not distinguish between data and programs, it can be used both as a (relational) data representation and as a query language.

Relational databases can be represented directly in PROLOG as a listing of all instantiated predicates corresponding to relation tuples. For instance, consider a small portion of the insurance database:

```
customer(3163, smith, 1935, london)
customer(3154, jones, 1942, atlanta)
    ...
covered _ by(3163, 'term insurance')
covered _ by(3163, 'whole life annuity')
    ...
```

While general relational schemas are defined only implicitly by the predicate names, PROLOG can be used as a powerful mechanism to define "generalized views" via additional rules. Views are used in DBMS to allow for more flexible data access. PROLOG rules differ from traditional view mechanisms in that with the use of variables they can accept parameters making them equivalent in this respect to the selector language construct proposed for database programming languages (Mall et al., 1983). Consider the following examples:

```
covered _ by _ many(Custom _ id) : —  covered _ by(Custom _ id, Benefit1)
                                      covered _ by(Custom _ id, Benefit2),
                                      not(Benefit1  =  Benefit2).

special _ customer(C _ id) : —  not(covered _ by _ many(C _ id))
                               or(customer(C _ id,N1,Y1,london),
                                  customer(C _ id,N2,Y2,paris)).
```

The first of these PROLOG statements can be read as: "A customer with customer id Custom_id (a variable) is covered by many benefits, *if* he/she is covered by at least two benefits which are different". Similarly, the second example reads: "A customer is special *if* he/she is not covered by many benefits and lives in Paris or London".

In addition to representing data and view definitions, PROLOG can also directly represent queries about base data or views. A query is simply a goal:

```
: —  covered _ by _ many(C _ id).
```

which when executed will return in C_id the customer id of a customer who is covered by many insurance benefits.

3.2.2. *Generalized Data Management*—a DBMS *Implemented in PRO-LOG.* A further step towards integrating the deductive capabilities of PROLOG with DBMS capabilities can be taken by implementing a general purpose DBMS directly in PROLOG—stage 2 in our approach. This can be done quite easily, and provides a means of adding flexible and general data access mechanisms to the inference engine without the need for a complicated interface to external database files.

In order to effect this stage in ES—DBMS coupling, the first requirement is the definition of an internal representation of a relational database. Given such a representation scheme, one can define any number of generalized operations to provide the facilities of a DBMS. Our approach (Vassiliou et al., 1983) provides a simple way to specify generalized relational operators acting on any relation and set of attributes. PROLOG programs map from this simpler, user-oriented view of the operations to their implementation for the particular database and representation scheme chosen. This provides a degree of logical data independence as in the traditional levelled architecture of DBMS.

A more application-specific solution is proposed in Kunifuji and Yokota (1982). Kowalski (1981) details the use of PROLOG for integrity constraints, database updates, and historical databases.

Another advantage of a generalized DBMS within PROLOG is efficiency. It is possible to devise a more sophisticated storage strategy (e.g., B–Trees), and perhaps to use auxiliary indexing schemes, hashing, etc. (Tarnlund, 1978). The work reported in Pereira and Porto (1982) demonstrates that, for specific applications, clever indexing schemes that guide decisions about which portions of external files should be read into the internal database can be devised. However, these strategies are not easily generalizable.

3.2.3. Loose Coupling of a PROLOG-based ES with a Relational DBMS. Conceptually the simplest solution to the problem of using existing databases is to extract a snapshot of the required data from the DBMS before the ES begins to work on a set of related problems—stage 3 of the approach. This portion of the database is stored in the internal database of the ES as described previously with PROLOG.

For this scenario to work, the following mechanisms are required:

1. Link to a DBMS with unload facilities;
2. Automatic generation of an ES database from the extracted database; and
3. Knowing in advance which portion of the database is required for extraction (static decision).

Usually, a superset of the actually required data will have to be extracted by this strategy. The approach may be rendered infeasible if this superset becomes too large to fit into the internal expert system DBMS.

3.2.4. Tight Coupling of a PROLOG-based ES with a Relational DBMS. For the fourth and final stage of our approach as has been implemented with PROLOG, we consider a very large existing database stored under a relational commercial DBMS. The naive use of the communication channel between the ES and the DBMS will assume the redirection of all ES queries, on predicates representing relations, to the DBMS for stored database relations. Any such approach is bound to face at least two major difficulties: the number of database calls will be prohibitively many (each PROLOG goal corresponds to a separate DBMS call), and the complexity of PROLOG goals (queries) may make it impossible to translate them directly to DBMS queries (e.g., recursion).

These difficulties can be overcome by collecting and jointly executing database calls rather than executing them separately whenever required by the ES. In essence, the pure depth-first approach of PROLOG is replaced by a combination of a depth-first reasoning and a breadth-first database call execution (Reiter, 1978).

In practice, we use an amalgamation of the ES language with its own meta-language, based on the 'reflection principle' (Weyhrauch, 1980). This allows for a deferred evaluation of predicates requiring database calls, while at the same time the inference engine (theorem prover) of the ES is working.

Since all inferences are performed at the meta-level in a simulation of object-level proofs, we are able to bring the complex ES queries into a form where some optimization and direct translation to a set of DBMS queries is feasible. Note, that at this point it would be desirable to have an intelligent DBMS that collectively optimizes the execution of all the queries. We shall discuss such ideas in section 4.3.

The queries are directed to the DBMS, answers are obtained and transformed to the format accepted by the ES for internal databases. The ES can continue its reasoning at the object level. Each invocation of predicates corresponding to database relations will now amount to an ES database goal, rather than a call to an external DBMS.

By tight coupling, we are now able to use an existing large relational database as an extension of any internal PROLOG database.

4. Experts for Database Systems

When compared to requirements for advanced business applications, such as decision support for managerial users, current database management systems display a number of weaknesses. In this section, we highlight these problems and identify some strategies for overcoming the limitations by the use of rule-based techniques. We shall not go into detail about other Artificial Intelligence approaches such as natural language interfaces (Harris, 1977; Wahlster, 1981) or conceptual modelling techniques (Brodie & Zilles, 1980; Brodie et al., 1983) which also contribute to better database design and usage.

4.1. Problems with Current Database Systems

With the advent of decision support systems and the proliferation of computers and databases in general, the target user population for database query languages has changed: there are more (potential) users, such as managers and application specialists who have a high degree of application knowledge but little patience to acquire much familiarity with programming concepts (Jarke & Vassiliou, 1982).

For such users, higher-level query languages are required which offer powerful operations without demanding major interaction skills. Besides the more ergonomic approach taken by the so-called second generation languages (Vassiliou & Jarke, 1984), reasoning capabilities embedded in the system may allow for a wider range and more concise formulation of queries.

A second problem arises if users want fairly complex operations to be performed on the data which are not provided by the query language. Currently, users are forced to use a database programming language (Schmidt et al., 1982) instead of an ad-hoc query language, and program the operations on the data explicitly. This is not easy for noncomputer specialists and often also prevents the use of standard DBMS report generation facilities, burdening the user with even more programming tasks.

A third difficulty with most current DBMS is the lack of support for operations performed in a *context*. The system does not have a partner model of the user (Wahlster, 1981). Neither does it recognize that multiple users working on the system at the same time ask queries that could be answered collectively (Jarke et al., 1982). Finally, a user who wishes to issue a query that is based on the result of a previous query has to store that result explicitly.

Besides not being as user-supportive as they could be, database systems could be improved towards more safety of the data and more efficiency of the execution of read and write transactions.

Safety is guaranteed in a database by enforcing integrity constraints, and by maintaining consistency through concurrency control. However, many DBMS support only rather simple types of integrity constraints because it is hard to test more complex combinations of constraints efficiently. Also, most systems do not support the resubmission of (sequences of) transactions after failure or user errors (Gray, 1981).

Efficiency mainly refers to the response time for evaluating queries. Many systems do not recognize all special cases of query structure or content that permit the use of efficient special-purpose algorithms. Neither do they recognize sequences of queries where the output of one can be used as the input of the next (e.g., recursion, focusing, etc.).

In the next two subsections, we explore how rule-based technology can be used to improve on these problems. As in section 3, we take a stagewise approach to the problem going from the user interface of the system towards its internal operations. It will be seen that many of the problems are related to each other and allow for common solution strategies.

Again, as in section 3, the question arises whether these strategies can be used only in newly developed database systems, or whether an external expert system can be used via a bridge on top of an existing database system. We shall not address this question in this section but postpone it to section 5 where we discuss general issues common to both directions of interaction.

4.2. Expert Application 1: Intelligent Database Usage

In the past few years, much research has been directed towards making database interfaces more intelligent. The main thrust is towards more deductive capabilities; less is known about enhanced functions to be applied to retrieved data.

The first DBMS allowed access only to the stored files, records, and fields (relations, tuples, and attribute values in the relational model of data [Codd, 1970]). More recent systems provide view mechanisms that allow the user to name windows through which only a subset of the database is visible. In the relational model, views are often called virtual relations. For example, a virtual relation "whole life insurance bearers" can be defined that contains customers who have a whole life insurance policy.

Traditional view mechanisms are somewhat limited. In the above example, a separate view of customers would have to be defined for each type of insurance benefit. The *selector* concept introduced in Mall et al. (1983) allows the definition of views with parameters.

In our example, a selector

with _ benefit (T) for customer _ relation

can be defined where T can assume values such as "whole life insurance" or "term annuity". The user can now request or alter the value of a selected variable

customer [with _ benefit('term annuity')]

which defines a moving window whose focus depends on the parameter chosen.

For querying purposes, PROLOG offers similar facilities in the "generalized view" rules presented in section 3. The PROLOG inference engine can perform deductions on these view definitions, and thus answer queries not immediately answerable from the stored data.

A number of researchers have investigated these possibilities and some implemented systems exist. The solutions can be classified in much the same way as the stages given in section 3. There are: integrated systems containing both the ES and the DBMS (Minker, 1978; Warren, 1981); loosely coupled systems where the ES does all its work before the DBMS is called (Chang, 1978; Fishman & Naqvi, 1982; Grishman, 1978; Henschen & Naqvi, 1982; Reiter, 1978); and finally the tightly coupled system DADM (Kellogg, 1982) that interleaves deduction in the ES with partial search in the DBMS. Besides offering the user a deductive formal query language, such systems can also support other high-level user interfaces such as natural language (Sagalowicz, 1977).

Other approaches which additionally address intelligent updating mechanisms will be described in section 4.3.2.

4.3. Expert Application 2: Intelligent Database Operation

The addition of general rules and inference mechanisms to a database system may improve the user interface, but it also presents a challenge to the database implementation researcher. Not only is the goal to execute deductive processing efficiently, but the new approach can be exploited as well to improve the execution of more traditional database operations. We investigate two areas: deduction-based query optimization methods, and deduction-based integrity checking for update transactions.

4.3.1. Query Optimization. It is of foremost importance to a DBMS that it contains a query evaluation subsystem which identifies a fast way to execute any submitted query. For this purpose, a query is usually standardized, simplified, and transformed in a way that makes the application of fast special-purpose algorithms feasible (Jarke & Koch 1982). Simplification and standardization can be quite complex if difficult queries are asked.

A deductive component may use metarules that guide the choice among the many applicable query transformation rules. Grishman (1978) and Reiter (1978) describe applications of deduction to the simplification step. Warren (1981) implements the well-known query transformation heuristics of testing sharp restrictions first, and separating detachable subqueries in logic.

Jarke and Koch (1983) developed a generalized heuristic called range nesting that combines both ideas. The deduction mechanism, however, is deterministic and contained in the query language compiler. An extension to this mechanism will use metarules to generate alternative strategies and compare their evaluation costs based on knowledge about storage structures and database statistics.

Another transformation strategy that uses deduction makes more direct use of the general rules that define the database intension (Reiter, 1978). Semantic query processing (Hammer & Zdonik, 1980; King, 1979, 1981) applies the integrity constraints of the database to simplify the execution of queries.

Assume, for example, that the ES contains a legal constraint that a minor may not get more than \$3000 annuity income. If now the database is queried for customers with annuities of more than \$3000 (or \$5000, etc.), minors can be excluded from the search. Depending on the physical database structure, this may or may not speed up execution. Thus, this strategy also needs metarules to guide the application of constraints.

Finally, the simultaneous optimization of multiple queries (Jarke et al., 1982) can be supported by an ES. One strategy is to remember selected query results to be used later (Finkelstein, 1982), another to recognize common subexpressions in a batch of queries. The execution of deductive queries, as outlined in the section on intelligent database usage, requires a sequence of related queries to be submitted to the database. The relationship between these queries can be exploited for optimization (Grant & Minker, 1981). A last area where expert knowledge is helpful is the complex problem of selecting physical storage structures and access paths (Paige, 1982).

4.3.2. Transaction Management.   Changes in the database must obey the general laws defined by static (i.e., data value, referential) and dynamic (i.e., data value change) integrity constraints. The DBMS has to determine which of the many general laws apply to a specific operation on specific data. Again, a pattern-matching oriented deduction process can be used (Henschen et al., 1982; Nicolas & Yazdanian, 1978). Metarules can be used to determine the possible delay of integrity checking until the changed data is actually needed (Lafue, 1982). While this idea may improve system efficiency, it is unclear whether it is feasible from a management standpoint to keep the integrity of the database unknown over an extended period of time.

A second question also affects the user interface of the system: how to react to violations? Nicolas and Yazdanian (1978) see three alternatives: reject the operation; accept the operation but do not change the database until further operations are submitted so that the transaction as a whole leaves the database in a consistent state (Gray, 1981); or trigger automatic changes to other data items to bring the database back to a consistent state.

The appropriate choice depends on many factors which can be partially controlled by metarules and partially by the user. A knowledge-based database architecture proposed in Jarke and Shalev (1984) contains an input management system to avoid unnecessary resubmission of transactions in such cases.

Language constructs such as the selector mechanism described in section 4.2. can be used for both query optimization and transaction control. First, the definition predicate of the selector can be used to identify the applicable integrity constraints for semantic query optimization and update control. Second, the selector defines candidate physical access paths which may lend themselves to "view indexing" (Roussopoulos, 1982), in which the collection of answers to the parameterized selection predicate is stored as a physical access path. Paige (1982) describes a method of "finite differencing" which dynamically generates and maintains views for query optimization and integrity control purposes.

5. Architectural and Technical Issues in Coupling

To summarize our overview of ES applications to database management, it can be said that the techniques used in the two proposed stages of intelligent use and intelligent operation are very similar. A system that incorporates most of the proposed strategies in a uniform architecture should be clearly feasible. On the other hand, such systems would need the same coupling mechanisms as the ones described in section 3 which in turn might greatly benefit from more intelligent databases. Therefore, in this last section of the paper we pose some architectural questions looking at the interaction between ES and DBMS from a higher perspective.

One might think that the natural architecture would be a uniform integrated system written in and usable through one language: either an extended database programming language (Schmidt et al., 1982) with deductive capabilities, or an ES language with general programming capabilities, such as PROLOG (Kowalski, 1981; Walker, 1984).

However, not only has each of these languages its own idiosyncrasies making it awkward to use in certain subsystems (e.g., complex computations in PROLOG), but this architecture also defeats the whole purpose of exploiting existing DBMS and quantitative methods. We therefore have to exclude this theoretically elegant alternative.

Three candidate architectures have been identified for the coupling of independent systems. This categorization is based on where processing takes place and how the interaction is controlled. We discuss each architecture in turn.

One architecture calls for a total distribution of processing and control. The two systems interact by exchanging messages, as in an "actor" approach (Hewitt, 1976; Dhar, 1983). Each interaction assumes a master to slave relationship between the originator and the receiver of the message, but both systems are self-contained and can be operated independently. An advantage of this architecture is a large degree of application and system independence, allowing for transportability to other ES and DBMS. How much each system has to know about the other's capabilities is an important consideration. The duplication of knowledge representations may introduce the usual dangers of redundancy: inconsistency and incompatibility.

At the other extreme, concentration of processing and control, a system integration can be envisioned. One of the two subsystems (ES or DBMS) may assume a more dominant role. This approach has naturally been followed by most researchers who focus on one direction of the interaction between ES and DBMS, for example Chang (1978), Fordyce and Sullivan (1983), and Kellogg (1982). The architecture suggests a more variable distribution of labor (e.g., where query optimization is done) than the typically predetermined separation of labor in distributed systems. There is much flexibility and potential in such an architecture, at the expense of transportability. Another difficulty with this solution is the integration of additional external subsystems.

Finally, in a third architecture, processing is distributed but control is now the responsibility of a separate subsystem, a supervisor program. In essence, the supervisor performs all the necessary steps for interfacing

the ES with the DBMS (e.g., translations), and manages the interaction between them. This appears as a compromise architecture with the main
advantage of allowing for a smoother interaction with other subsystems.
Such subsystems include mathematical models, and modules for knowledge
acquisition, generation of alternatives solutions (Reitman, 1982), and user
interfaces. The challenging research question would be how to implement
a supervisor that makes full use of the capabilities of the various subsystems
without duplicating their features.

Within each architecture, means must be provided to translate knowledge representations and transaction requests between subsystems. As
pointed out in section 3, the similarity between PROLOG and the relational
data model makes this task relatively easy. If other database models are
used, however, intermediate translation and optimization steps become
necessary. Depending on the architecture chosen, it must be decided which
subsystem is responsible for this task.

6. Concluding Remarks

Several practical benefits from the cooperative use of expert systems with
database management systems were identified. The overall goal of our work
is an advanced system effectively supporting business decisions. Such a
system would integrate database management, model management, and
expert system technology within the same architecture.

At New York University, we are exploring several strategies for achieving this goal. A stage-wise approach requiring simultaneous research on
complementary topics as outlined in this paper is being designed and implemented.

7. References

Bowen, K. A., & Kowalski, R. A. Amalgamating language and metalanguage in logic programming.
    In K. Clark & S. A. Tarnlund (Eds.), *Logic programming*. New York: Academic Press, 1982.
Brachman, R. On the epistemological status of semantic networks. In N. V. Findler (Ed.), *As-
    sociative networks: Representation and use of knowledge by computer*. New York: Academic
    Press, 1977.
Brodie, M., Zilles, S. (Eds.) Proceedings workshop on data abstraction, databases, and con-
    ceptual modelling. *SIGMOD Record*, 1980, *11*(2).
Brodie, M., Mylopoulos, J. & Schmidt, J. W. (Eds.). *Perspectives on conceptual modelling*. New
    York: Springer, 1983.
Chandrasekaran, B. Expert systems: Matching techniques to tasks. In W. Reitman (Ed.), *Artificial
    intelligence applications for business*. Norwood, NJ: Ablex, 1984.
Chang, C. L. DEDUCE 2: Further investigations of deduction in relational databases. In H.
    Gallaire & J. Minker (Eds.), *Logic and databases*. New York: Plenum, 1978.
Clocksin, W. F., & Mellish, C. S. *Programming in PROLOG*. New York: Springer, 1981.

Codd, E. F. A relational model for large shared data banks. *CACM,* 1970, *13*(6), 377–387.

Davis, R. Expert systems: Where are we? And where do we go from here? *Proceedings of the Seventh International Joint Conference on Artificial Intelligence,* 1981.

Dhar, V. *Designing an intelligent decision support system for long range planning: An artificial intelligence approach.* Unpublished doctoral dissertation, University of Pittsburgh, 1983.

Finkelstein, S. Common expression analysis in database applications. *Proceedings of the ACM-SIGMOD conference,* 1982, 235–245.

Fishman D., & Naqvi, S. An intelligent database system: AIDS. *Proceedings Workshop on Logical Bases for Data Bases,* 1982.

Fordyce, K. J., & Sullivan, G. A. *Artificial intelligence in decision support systems: Presentation of an AI technology and its application in working DSS.* Unpublished manuscript, Poughkeepsie, 1983.

Gallaire, H., & Minker, J. (Eds.). *Logic and databases.* New York: Plenum, 1978.

Grant, J., & Minker, J. Optimization in deductive and conventional relational database systems. In H. Gallaire, J. Minker, & J. M. Nicolas (Eds.), *Advances in database theory.* New York: Plenum, 1981.

Gray, J. The transaction concept: Virtues and limitations. *Proceedings of the Seventh VLDB Conference,* 1981, 144–154.

Grishman, R. The simplification of retrieval requests generated by question answering systems. *Proceedings of the Fourth VLDB Conference,* 1978, 400–406.

Grosz, B. TEAM extended abstract. In P. Buneman (Ed.), *Proceedings of the Philadelphia Database Interface Workshop,* 1982.

Hammer, M., & Zdonik, S. Knowledge-based query processing. *Proceedings of the Sixth VLDB Conference,* 1980, 137–147.

Harris, L. R. User-oriented data base query with the ROBOT natural language query system. *Proceedings of the Third VLDB Conference,* 1977, 303–312.

Henschen, L., McCune, W., & Naqvi, S. Compiling constraint-checking formulas from first-order formulas. *Proceedings of the Workshop on Logical Bases for Data Bases,* 1982.

Henschen, L., & Naqvi, S. On compiling queries in recursive first-order databases. *Proceedings of the Workshop on Logical Bases for Data Bases,* 1982.

Hewitt, C. *Viewing control structures as patterns of passing messages* (AI Memo 410). Cambridge, MA: MIT, December 1976.

Jarke, M., & Koch, J. A survey of query optimization in centralized database systems (NYU Working Paper Series, CRIS #44, GBA 82–73 (CR)) New York: New York University, November 1982.

Jarke, M., & Koch, J. Range nesting — A fast method to evaluate quantified queries. *Proceedings of the ACM–SIGMOD Conference,* 1983, 196–206.

Jarke, M., Koch, J., Mall, M., & Schmidt, J. W. Query optimization research in the database programming languages (DBPL) project. *Database Engineering,* 1982, *5,* 11–14

Jarke, M. & Shalev, J. A database architecture for supporting business transactions. *Journal of Management Information Systems* 1984, *1*(1).

Jarke, M., & Vassiliou, Y. Choosing a database query language. Manuscript submitted for publication, 1982.

Kellogg, C. Knowledge management: A practical amalgam of knowledge and data base technology. *Proceedings of the Third National Conference on Artificial Intelligence,* 1982.

King, J. J. *Exploring the use of domain knowledge for query processing efficiency* (Tech. Rep. STAN–CS–79–781). Stanford University, December 1979.

King, J. J. QUIST: A system for semantic query optimization in relational data bases. *Proceedings of the Seventh VLDB Conference,* 1981, 510–517.

Kowalski, R. Logic as a database language. Unpublished manuscript, Imperial College, London, 1981.

Kunifuji, S., & Yokota, H. Prolog and relational databases for fifth generation computer systems. *Proceedings Workshop on Logical Bases for Data Bases,* 1982.

Lafue, G. Logical foundations for delayed integrity checking. *Proceedings Workshop on Logical Bases for Data Bases,* 1982.

Mall, M., Reimer, M., & Schmidt, J. W. Data selection, sharing, and access control in a relational scenario. In M. Brodie, J. Mylopoulos, & J. W. Schmidt (Eds.), *Perspectives on conceptual modelling.* New York: Springer, 1983.

Minker, J. An experimental relational database system based on logic. In H. Gallaire & J. Minker (Eds.) *Logic and Databases.* New York: Plenum, 1978.

Minsky, M. A framework for representing knowledge. In P. H. Winston (Ed.), *The psychology of computer vision.* New York: McGraw-Hill, 1975.

Naqvi, S., Fishman, D., & Henschen, L. An improved compiling technique for first-order databases. *Proceedings Workshop on Logical Bases for Data Bases,* 1982.

Nau, D. Expert computer systems. *Computer,* February 1983, 63–85.

Nicolas, J.-M, & Yazdanian, K. Integrity checking in deductive databases. In H. Gallaire & J. Minker (Eds.), *Logic and databases.* New York: Plenum, 1978.

Nicolas, J.-M, & Demolombe, R. On the stability of relational queries. *Proceedings Workshop on Logical Bases for Data Bases,* 1982.

Olson, J. P., & Ellis, S. P. PROBWELL — An expert advisor for determining problems with producing wells. *Proceedings of the IBM Scientific/Engineering Conference,* 1982, 95–101.

Paige, R. Applications of finite differencing to database integrity control and query/transaction optimization. *Proceedings Workshop on Logical Bases for Data Bases,* 1982.

Pereira, L. M., & Porto, A. A PROLOG implementation of a large system on a small machine. Departmento de Informatica, Universidade Nova de Lisboa, 1982.

Reiter, R. Deductive question-answering on relational data bases. In H. Gallaire & J. Minker (Eds.), *Logic and Databases.* New York: Plenum, 1978.

Reitman, W. Applying artificial intelligence to decision support: where do the good alternatives come from? In M. J. Ginzberg, W. Reitman, & E. A. Stohr (Eds.), *Decision Support Systems.* New York: Elsevier North-Holland, 1982.

Robinson, J. A. A machine oriented logic based on the resolution principle. *JACM, 1*(4), 1965, 23–41.

Roussopoulos, N. View indexing in relational databases. *ACM-TODS,* 1982, *7*(2), 258–290.

Sagalowicz, D. IDA: An intelligent data access system. *Proceedings of the Third VLDB Conference,* 1977, 293–302.

Schmidt, J. W., Mall, M., Koch, J., & Jarke, M. Database programming languages. In P. Buneman (Ed.), *Proceedings of the Philadelphia Database Interface Workshop,* 1982.

Stohr, E. A., & White, N. H. User interfaces for decision support systems: An overview. *International Journal of Policy Analysis and Information Systems,* 1982, *6*(4), 393–423.

Schank, R. C. *Conceptual information processing.* New York: Elsevier North-Holland, 1975.

Tarnlund, S. A. *Logical Basis for Data Bases.* Unpublished manuscript, 1978.

Travis, L, & Kellogg, C. Deductive power in knowledge management systems: Ideas and experiments. *Proceedings Workshop on Logical Bases for Data Bases,* 1982.

Tsichritzis, D. Form Management. *CACM,* 1982, *25,* 453–478.

Vassiliou, Y., Clifford, J., & Jarke, M. *How does an expert system get its data?* (NYU Working Paper CRIS #50, GBA 83–26 (CR)). (extended abstract in *Proceedings of the Ninth VLDB Conference,* 1983. V, 70–72)

Vassiliou, Y., & Jarke, M. Query languages — a taxonomy. In Y. Vassiliou (Ed.), *Human factors and interactive computer systems.* Norwood, NJ: Ablex, 1984.

Wahlster, W. Natural language AI systems: State of the art and research perspective. In J. Siekmann (Ed.), *Proceedings of the German Workshop on Artificial Intelligence 81.* New York: Springer, 1981.

Walker, A. Data bases, expert systems, and PROLOG. In W. Reitman (Ed.), *Artificial intelligence applications for business.* Norwood, NJ: Ablex, 1984.

Warren, D. H. D. Efficient processing of interactive relational data base queries expressed in logic. *Proceedings of the Seventh VLDB Conference,* 1981, 272–282.

Warren, D. H. D., Pereira, L. M., & Pereira, F. PROLOG — the language and its implementation compared with LISP. *(Proceedings of the Symposium on Artificial Intelligence and programming languages.) SIGPLAN Notices,* 1977, *12*(8), 109–115.

Waterman, D., & Hayes-Roth, F. (Eds.). *Pattern directed inference systems.* New York: Academic Press, 1979.

Weyhrauch, R. Prolegomena to a theory of mechanical formal reasoning. *Artificial Intelligence,* 1980, *13,* 133–170.

Databases, Expert Systems, and PROLOG

Adrian Walker
IBM RESEARCH LABORATORY K51
5600 COTTLE ROAD
SAN JOSE, CA 95123

The knowledge needed for an expert system is usually held in the form of facts and IF-THEN rules. The expert system itself is a program (often in LISP) which infers advice from the facts and the rules.

In the language PROLOG, a program consists of facts and rules, so the knowledge needed for an expert system can be written down directly. To get advice, one can either use PROLOG's built in inference mechanism, or one can write further rules about how to use the knowledge. The facts in PROLOG correspond to a relational database, while the rules contain expertise about how to use the data.

This chapter uses specific examples, taken from three programs, to show that PROLOG is a practical language for bringing together several apparently diverse techniques in databases and expert systems.

1. Introduction

In the early days of computers in business, the most useful programs were those which carried out some well-understood well-structured activity, such as payroll computation. Each time the steps of some business activity were well enough understood to be written down as a flowchart, a new program could be written to handle the activity.

Starting around the early 1960s, it was noticed that many programs shared data with other programs. Since each program used data in its own way, there was a tendency to work with overlapping data in different formats. A change of data format could well result in the need to rewrite a large number of programs. It became clear that much effort could be saved, and services could be improved, if formats and data were factored out of programs and into a common dictionary and database. A distillation of the access methods and sharing protocols which had been scattered across various programs was concentrated in a single gateway to the data, the DBMS, or database management system.

Once the data are in one place, and are accessed by standardized methods and with standardized controls, the development of new application programs is considerably easier (Codd, 1982). The data are easy to change, and to share amongst programs. The DBMS may provide services, such as views (IBM, 1981; Selinger, Astrahan, Chamberlin, Lorie, & Price, 1979), to insulate programs from changes in data formats. The cost of the DBMS can be spread over many applications.

However, even with the rationalization of data handling provided by DBMS's, the demand for new application programs seems to outstrip what can be supplied using present software technology. Each business task must be well-structured and understood before it can be programmed and made useful to an organization. Existing application programs can often be adapted to new uses only with considerable effort. Much expertise is needed to match a task to existing software, and to specify new software to fill in any gaps.

By the early 1970s, it was noticed that much of the difficulty in structuring new tasks comes from two sources. The first is that, while a person may be expert at solving problems in some domain, he or she may not be able to describe a step-by-step procedure in sufficient detail for a computer professional to be able to write down a useful flowchart. The second source of difficulty is related to the first. Until recently, it has not been possible to just describe to a computer what we want it to do. We have also had to write down, in sometimes excruciating detail, the steps which must be followed to get the end result that we want.

A major step towards a solution to the difficulty of structuring tasks for the computer came from a person who is both an expert in a task domain (medicine) and highly qualified in computing. Shortliffe (1976) proposed that the knowledge that a person uses to solve a problem can usefully be written down as a collection of assertions and IF-THEN rules. Each assertion or rule in the collection can have a confidence factor, indicating the degree of belief we have in its validity.

This way of writing down knowledge about a task was used with some success in the MYCIN program for medical diagnosis. Broadly, MYCIN consists of (a) a collection of rules and assertions, and (b) an inference engine which puts the assertions and rules to work to make a diagnosis.

Initial success with MYCIN led people to apply the program to different tasks (Van Melle, Scott, Bennett, & Peairs, 1981), and thereby to find out which parts of the program had to be changed, and which could remain the same, when switching to a new task. It became clear that, just as a DBMS remains about the same while handling different databases, the MYCIN inference engine could stay about the same while handling some different collections of assertions and rules. The factoring out of task-specific knowledge led to the EMYCIN program (van Melle, et al., 1981), an inference engine which has been applied to medical (Aikins, Kunz, Shortliffe, & Fallat, 1982) and even nonmedical tasks by providing different collections of rules.

Perhaps the most important aspect of EMYCIN is that it is not necessary, when using it, to specify a step-by-step procedure for a task. We do not have to say how to diagnose a patient. We just provide some rules about medicine, and we say what we want done, namely a diagnosis. To the extent that EMYCIN makes this possible, it represents a valuable step towards bridging the gap between a human expert and a computer. Writing down rules seems to be much more productive than writing down flowcharts. Similar observations underlie the design of some other programs. The OPS–5 inference engine (Forgy, 1981) has been used with rules for configuring computer installations. SYLLOG (Walker, 1981) is a language in which one can gradually accumulate knowledge, expressed as English-like syllogisms, about a task domain; the knowledge can be used to access a relational database.

In an EMYCIN application, the results of two quite different styles of programming are used. The inference engine is programmed in the language LISP, while the rules about a task domain can be provided by a person who may have little programming experience. If the EMYCIN inference engine is actually suitable for the task domain, then all goes smoothly. However, as new tasks further from medical diagnosis are tried, it is often necessary to change the behavior of the inference engine (Ennis, 1982). One can either go into the LISP program and change it, or one can specify extra rules about how to use the rules differently. In the first case, there is a feeling of going back to square one. We have to specify step-by-step how the computation is to proceed, rather than just writing down what we want to happen. In the second case, it is not always easy to keep a clear separation between the extra rules that are needed to make correct inferences and the actual rules for the task at hand. This can create problems if we wish to give explanations of the advice produced by the system.

Since IF-THEN rules seem to be a good way of writing down what we want computed, rather than how we want it computed, why not just specify an inference engine as a collection of IF-THEN rules? We should then be able to change the way in which deductions are made by altering a rule or two, rather than by rewriting parts of a LISP program or changing our collection of task-specific rules. If this can be done, it will factor out both the inference engine and the task rules, into a rule notation which is useable by people who are not professional programmers.

It turns out that the language PROLOG (Clocksin & Mellish, 1982; Kowalski, 1979), allows us to do just this. A program in PROLOG consists of facts and rules. If we have a collection of rules about a task, we can either use PROLOG's built-in inference mechanism to get advice, or we can write further rules about what to do with the collection. These further rules amount to a specification of what we want an inference engine to do.

In practice, there are several advantages in adopting the PROLOG approach just outlined. Both the task rules and the rules specifying the inference engine are in the same language. A sentence of the language is of the nature "what I want done is . . .", rather than "here are all the steps which must be followed . . .". Moreover, some EMYCIN-like inference engines can be specified in just a few rules—sometimes as few as four. These rules can be quite clearly separated from the usually large number of task rules. The total meaning of the task rules can then be changed at will by minor tuning of the rules for the inference engine.

Reviewing our informal sketch of some applications of computers to problems of interest to business, we see that, in the early days, each program was relatively task-specific. Next, data were factored out of programs into databases, the DBMS software spanned many tasks, but each task needed a different application program to drive the DBMS. It was then noticed that there were difficulties in transferring expertise from people to step-by-step application programs. EMYCIN-style programming promised to ease the situation by providing a fixed inference engine which could compute using different sets of IF-THEN rules for different tasks. However, for some tasks it was after all necessary to reprogram the inference engine. So we were led to the idea of specifying not only the task, but also the inference engine, using IF-THEN rules. The language PROLOG supports this idea in practice.

Aside from PROLOG's chameleon-like ability to change its style of making deductions by the addition of rules describing an inference engine, several concepts which are central to databases and expert systems are encapsulated very clearly. Section 2 of this paper describes a knowledge based program which can be used for access to relational databases with views. Section 3 shows how the IF-THEN rules of expert systems fit naturally in PROLOG. Section 4 describes the fact that the syntax and semantics of a natural language such as English (McCord, 1982; Pereira, Sabatier & Oliveira, 1982; Walker & Porto, 1982; Warren & Pereira, 1981) fit naturally in PROLOG. For example, for simple experiments, it is enough to write down rules for the desired syntax and semantics; there is no need to write a parser. Finally, section 5 consists of conclusions.

2. Data Bases

This section describes databases with views, and shows how these can be extended, by storing rules as well as data, to become knowledge-based systems. Section 3 will then carry the extension further to expert systems in which judgemental, as well as certain knowledge, is stored. Then, section 4 will describe a way of questioning a knowledge base in English.

Consider a chain of stores selling house and garden products, and suppose that the stock levels for a store are to be kept in a relational database. The kinds of facts to be kept are:

the stock level of product A is 25

product A should be reordered if stock below 30

product A is used to kill weeds

Views of the data, such as

reorder products which kill weeds

are also to be supported.

One way to describe a relational database with views is to use the language SYLLOG (Walker, 1981), which can be thought of as a means by which nonprogrammers can use the language PROLOG. To set up the stock level database in SYLLOG, one would start with a blank screen and type in

the stock level of product eg _ name is eg _ quantity

A	25
B	300
C	227
D	15
E	98
F	30

The 'eg _' in front of eg _ name and eg _ quantity in the first line indicates that they are example elements, as in Zloof (1977), which can stand for any product name and any quantity. So the first line forms a heading for a table with two columns, eg _ name and eg _ quantity. The line of dashes just separates the heading from the data in the body of the table.

Similarly, one can get another blank screen, and type in

product eg _ name should be reordered if stock below eg _ number
----

A	30
B	50
C	100
D	55
E	100
F	50

If one now asks for a prompt about what to do next, SYLLOG puts on the screen the invitation to

> make a command using these and other sentences:
>
> the stock level of product eg _ name is eg _ quantity
> product eg _ name should be reordered if stock below eg _ number

To retrieve either of the above tables, one just leaves the heading sentence on the screen, underlines it, and pushes the enter key. One can also make more specific retrievals, e.g., by changing an example element in a heading to an actual item. For instance, if one changes the prompt screen above to

> the stock level of product A is eg _ quantity
> ---

and pushes enter, then the screen changes to

> the stock level of Product A is eg _ quantity
> ---
> 25

So far, SYLLOG has been used simply to enter data and to retrieve it. Data can also be changed on the screen after it has been retrieved. Further, views of the data can be made known to SYLLOG by means of rules. For example, starting with the current prompt

> make a command using these and other sentences:
>
> the stock level of product eg _ name is eg _ quantity
> product eg _ name should be reordered if stock below eg _ number

on the screen, one can change the screen to

> the stock level of product eg _ name is eg _ quantity
> product eg _ name should be reordered if stock below eg _ number
> eg _ number greater than eg _ quantity
> ---
> reorder product eg   name

and push enter. The system now knows a rule for reordering products. The rule can be read as "if the stock level of product eg _ name is eg _ quantity, and product eg _ name should be reordered if stock below eg _ number, and eg _ number greater than eg _ quantity, then reorder product eg _ name".

When the rule has been given to SYLLOG, the next prompt is

make a command using these and other sentences:

the stock level of product eg _ name is eg _ quantity
product eg _ name should be reordered if stock below eg _ number
reorder product eg _ name

If one now leaves just the heading

reorder product eg _ name
----

on the screen, then the products which should be reordered appear like this

reorder product eg _ name

                    A
                    D
                    E
                    F

If one wishes to see the rule about which products should be reordered, one can get the prompt on the screen and leave just

reorder product eg _ name

The rule

the stock level of product eg _ name is eg _ quantity
product eg _ name should be reordered if stock below eg _ number
eg _ number greater than eg _ quantity
----
reorder product eg _ name

then appears. It can be changed on the screen, just as though it were data, if necessary.

So far, two tables of data and a rule for using the data have been put into the SYLLOG system. One might want to have an additional rule about how many units of each product to reorder. The above rule can be copied to an empty part of the screen, and the copy can be modified to

the stock level of product eg _ name is eg _ quantity
product eg _ name should be reordered if stock below eg _ number
eg _ number greater than eg _ quantity
eg _ number − eg _ quantity = eg _ short
eg _ short + eg _ number = eg _ required
----
reorder eg _ required units of product eg _ name

This is a new rule which expresses one possible reordering policy, namely: bring the stock of each item which is below threshold to twice the threshold. If one leaves the two rules on the screen, and pushes enter, then the next prompt is

> make a command using these and other sentences:
>
> the stock level of product eg _ name is eg _ quantity
> product eg _ name should be reordered if stock below eg _ number
> reorder product eg _ name
> reorder eg _ required units of products eg _ name

The system has absorbed the new rule, which can be used by leaving the heading

> reorder eg _ required units of product eg _ name
> --

on the screen and pushing enter. The amounts to be reordered appear below the heading as

> reorder eg _ required units of product eg _ name
> --

35	A
95	D
102	E
70	F

The system has so far acquired two tables of data and two rules containing knowledge about how to use the data. Rules need not work directly on the data, but may work on it indirectly through other rules. One can insert another table of data

> product eg _ name is used to eg _ treat eg _ items
> --

A	kill	weeds
B	kill	pests
C	kill	pests
D	kill	insects
E	kill	insects
F	kill	weeds
F	fertilize	lawns

and another rule

> product eg _ name is used to eg _ treat eg _ items
> reorder product eg _ name
> --
> reorder products which eg _ treat eg _ items

If one now gets the prompt, then selects the heading

```
reorder products which eg _ treat eg _ items
------------------------------------------------------
```

and presses enter, then

```
reorder products which eg _ treat eg _ items
------------------------------------------------------
                      kill        insects
                      kill        weeds
                      fertilize   lawns
```

appears on the screen. In making this retrieval, the system has chained two rules together, and has used all three tables of data. The database has been accessed via the knowledge contained in the rules.

   If the answer to a question is just 'yes' or 'no', rather than a table of results, then SYLLOG automatically supplies an explanation of the reasoning it used to get the answer. If one asks whether products which kill weeds should be reordered, by placing

```
reorder products which kill weeds
-------------------------------------------
```

on the screen, then the answer is

```
reorder products which kill weeds
-------------------------------------------

Yes, that's true

Because . . .

product A is used to kill weeds
reorder product A
-------------------------------------------
reorder products which kill weeds

the stock level of product A is 25
product A should be reordered if stock below 30
30 greater than 25
---------------------------------------------------------
reorder product A
```

On the other hand, if one asks whether products which kill pests should be reordered, by placing

```
reorder products which kill pests
-------------------------------------------
```

on the screen, then the answer is

```
reorder products which kill pests
----------------------------------------
Sorry, no
Because . . .
product B is used to kill pests
reorder product B
-----------------------------------------------------
reorder products which kill pests

the stock level of product B is 300
product B should be reordered if stock below 50
50 greater than 300 ?
----------------------------------------------------------
reorder product B
```

The explanation shows that the answer to the question is 'no' because there is enough of product B in stock, as indicated by the question mark in the third sentence of the last syllogism.

If one asks a question about something that is not in the knowledge base, then an explanation can help to point out the problem. For example, the question "should products which fertilize weeds be reordered?" can be asked by putting

```
reorder products which fertilize weeds
---------------------------------------------
```

on the screen. Then answer is

```
reorder products which fertizile weeds
---------------------------------------------
Sorry, no
Because . . .
product eg _ name is used to fertilize weeds?
```

In the explanation, the sentence with the question mark shows that there is no product for fertilizing weeds in the knowledge base.

Knowledge, expressed as rules, can be used both for retrieval, and to express integrity conditions or assertions for a database. For example, one could write a rule

```
product eg _ name should be reordered if stock below eg _ number1
product eg _ name should be reordered if stock below eg _ number2
eg _ number1 not equal eg _ number2
------------------------------------------------------------------------
error
```

to say that it is an error if, for some product, there are two different reordering thresholds.

One can also write collections of rules which have the effect of iterating over the data to make a retrieval. Such examples are given in Walker (1984). Essentially, what happens in these cases is that a rule makes use not only of other rules and of the data, but also refers back to itself.

In a conventional database management system, a user normally communicates with an application program which contains task-dependent knowledge that cannot easily be examined or changed. In database query languages for casual users one can make up queries on the screen, but the task-dependent knowledge stays mainly in the head of the user. The SYLLOG system can gradually acquire task dependent knowledge in the form of syllogism-like rules. In fact, the system checks incoming rules, and detects some possible sources of difficulty, such as rules which are special cases of other rules. Rules have the same status as data, that is, they can be freely examined and changed.

Although the SYLLOG language described above is English-like, it is not English. This is a disadvantage, in that one must learn a few conventions, such as the use of an 'eg' followed by an underscore for example elements. (We describe English access to a knowledge base in section 4). However, SYLLOG's lack of knowledge of English leaves the system quite flexible: rules can equally well contain phrases in French, German, or specialized business, legal or technical dialects. These phrases are introduced to the system at the time they are typed in. They then become part of the prompt, and can be used immediately.

SYLLOG satisfies one criterion for a knowledge-based system, namely, the knowledge is easy to examine, change, and extend. A second criterion, that the system should be able to explain its deductions, is also satisfied. A criterion for saying that a system is expert, (as well as knowledge-based), is that the system be able to reason judgementally (as well as exactly). The next section describes some rule-based expert system techniques.

3. Expert Systems

The last section described an approach to knowledge-based data management. This section shows a way in which the same approach can be extended in a MYCIN-style expert system which reasons judgementally.

Suppose that our store which sells house and garden products can call on an expert in plant diseases and their treatment. Over time, the expert builds a good reputation, and other stores in our chain would also like expert advice. However, there are now too many customers for the expert to deal with. We have a problem.

Our solution is to encode the expert's knowledge as a set of if-then rules, in a system which can be used, on a computer terminal, by customers or assistants in each store. Our expert system should at least be able to take over the routine questions, leaving our human expert to deal with the more important cases, and to keep the rule base up to date.

Experts often reason judgementally, and a question about a plant (e.g., is it wilting?) may not always have a clear cut yes or no answer. So we need a way of making judgemental inferences. In MYCIN style, we can extend the kind of knowledge rules used in the last section to attach a confidence factor, between 0 and 100, to each rule. For example, the rule

```
if 'plant is wilting' and
   'leaves have yellow spots' and
   not 'leaves are dropping' then 'there are mites' : 60.
```

says that if a plant is wilting and its leaves have yellow spots but are not dropping off, then it is probable (with confidence 60%) that there are mites on it.

In an experimental expert system called PLANTDOC, there are a number of such rules, together with an inference engine for extracting advice from the rules. Following MYCIN's style, PLANTDOC reasons like this. If the answers to the questions

```
plant is wilting?
leaves have yellow spots?
leaves are dropping?
```

are yes, yes, and no respectively, then the system concludes, with confidence 60%, that there are mites on the plant. (If the 60% confidence is not high enough, the system may also ask directly whether the cobwebs of mites can be seen.) However, if the answers to the first two questions are positive, but not definite, then one can type in numbers like 80 and 70 to so indicate. If one then types in 'no' for the last question, the system then takes the smaller of 80 and 70, and multiplies it by the confidence of 60% in the rule, to get a confidence of 42% that there are mites on the plant.

PLANTDOC also has rules about which products or treatments to use once it has estimated the cause of a problem. Most of these rules are certain, e.g.,

```
if 'mites' then 'apply product G' : 100
```

The certain and judgemental rules about symptoms, diseases, and treatments, together with the PLANTDOC inference engine (which is also a collection of PROLOG rules), support the following dialog on a terminal. In the dialog, the system poses questions. A question can be answered with 'yes', 'unk' (for unknown), 'no', or a number between −100 and 100 which indicates one's degree of belief that the answer is yes. A question can also be answered with 'why', causing the system to pause and explain its line of reasoning. When enough questions have been answered, the system then gives diagnoses, and recommended products or treatments.

```
plant is wilting? yes
leaves have yellow spots? yes
leaves are dropping? no
yellow leaves? no
leaves are curled? no
leaves have brown spots? no
my diagnoses are:
there are mites
it is probable (60%) that apply product G is suitable
there are leaf hoppers
it is possible (40%) that apply product F is suitable

again? yes
plant is wilting? 80
leaves have yellow spots? 70
leaves are dropping? no
yellow leaves? no
leaves are curled? no
leaves have brown spots? no
my diagnoses are:
there are mites
it is possible (42%) that apply product G is suitable
there are leaf hoppers
it is possible (28%) that apply product F is suitable

again? yes
plant is wilting? non
please say why, yes, unk, no, or number between −100 and 100
plant is wilting? no
yellow leaves? yes
top leaves went yellow first? why
(i.e. why is it important whether: top leaves went yellow first?)
```

because . . .

if	yellow leaves
	top leaves went yellow first
and	veins in leaves are green
then	it is possible that: iron deficiency

continue (con) or more explanation (why)? con

top leaves went yellow first? yes

veins in leaves are green? yes

leaves are curled? no

leaves have brown spots? no

my diagnosis is:

iron deficiency
it is possible (30%) that apply product H is suitable

again? yes

plant is wilting? no

yellow leaves? no

leaves are curled? yes

leaves are discolored? why
(i.e. why is it important whether: leaves are discolored?)

because . . .

if	leaves are curled
	leaves are discolored
and	leaves are reduced in size
then	it is definite that: there are aphids

continue (con) or more explanation (why)? con

leaves are discolored? yes

leaves are reduced in size? 80

leaves have brown spots? no

my diagnosis is:

there are aphids
it is probable (64%) that apply product J is suitable

again? yes

plant is wilting? no

yellow leaves? no

leaves are curled? no

leaves have brown spots? yes

leaves are curling down? −60

leaves are dying? no

my diagnosis is:

there are fungi
it is probable (50%) that apply product M is suitable

again? yes

plant is wilting? no

yellow leaves? no

leaves are curled? yes

leaves are discolored? no

leaves have brown spots? yes

leaves are curling down? why

(i.e. why is it important whether: leaves are curling down?)

because . . .

if	leaves have brown spots
	not leaves are curling down
and	leaves are dying
then	it is definite that: there is too much salt

continue (con) or more explanation (why)? why
(i.e. why is it important whether: there is too much salt?)

because . . .

if	there is too much salt
then	it is definite that: improve drainage

leaves are curling down? −60

leaves are dying? 80

my diagnoses are:

there is too much salt
it is probable (64%) that improve drainage is suitable

there are fungi
it is probable (50%) that apply product M is suitable

The main goal of PLANTDOC is to find products or treatments which will solve the problem at hand. To do so, the system checks the products and treatments given in the rules, and asks questions at the terminal about items for which it has no rules.

In trying to show that a product or treatment is suitable, the system first estimates whether it is plausible. To do so, an internal inference is run. This internal inference uses any answers which have been given for previous questions, but asks no questions of its own. Rather, if an unanswered question is found, the answer is assumed (temporarily) to be yes. The result of the internal inference is a best case plausibility that the product or treatment is suitable. If this plausibility is high enough, the system actually begins to ask for information from the person at the terminal. Otherwise, the system estimates the plausibility of the next product. In this way, the questions which are asked are trimmed to those which could lead to a decision about a product.

This section has described the PLANTDOC experimental expert system. The rules and inference method in PLANTDOC are an extension, to get judgemental reasoning using confidence weights, of the yes-no rules of the SYLLOG system in section 2. As with many knowledge-based and expert systems, SYLLOG and PLANTDOC communicate on a terminal mainly by English sentences that are prestored. In SYLLOG, English-like sentences can be typed in, but not free English. The next section describes a system into which one types English questions about products in a garden store.

4. English Question Answering

The last two sections have described the SYLLOG and PLANTDOC systems, which carry on a dialog without processing English sentences in detail. In the case of SYLLOG, one can type in sentences, provided one follows certain conventions. In the case of PLANTDOC, the rules consist of English phrases that are combined for output, but English sentences describing a problem with a plant cannot be typed in.

This section describes Kbo1, a prototype knowledge-based system which answers English questions about some of the products supplied by a garden store (Walker & Porto, 1982). The system matches, to a certain extent, the behavior of a helpful, knowledgeable assistant in a store which sells products such as domestic pesticides and weed-killers.

Some questions which produce brief, but useful and accurate answers, are:

> what products do you sell?
>
> what is each product that you sell for?
>
> what can I use to kill weeds in my lawn in spring?
>
> what can I use to kill weeds around my fence?
>
> do I need a sprayer to use product A?
>
> what is the response time of weeds to product A?
>
> does product A kill dandelions in less than 20 days?

On an IBM mainframe computer, each question is answered less than one second.

The system only answers questions related to certain house and garden products. Questions which lie outside this scope, such as

> is there a bus stop near here?

are answered with a sentence such as

I'm sorry, I don't know the word: bus

To use the Kbo1 system one types English questions on a terminal, and gets answers on a screen. In the following example session, a question starts after a prompt symbol ">>"and ends with a question mark "?". The answer produced by the system follows the question. Comments which we have added appear in parentheses after answers.

>> what products do you sell?

product A
product B
product C
product D
product E
product F

(the system has simply listed the products.
for examples with more products, it would print just the kinds of products)

>> what do you know about?

product A kill weed
product B kill pest
product C kill pest
product D kill insect
product E kill insect
product F kill weed
product F fertilize lawn

(the system has listed the products in its knowledge base, with one or more intended uses for each product. English output generation is minimal)

>> what is each product that you sell for?

product A kill annualweed
product A kill perennialweed

product B kill gardenpest

product C kill gardenpest

product D kill crawlinginsect
product D kill flyinginsect
product D kill hoppinginsect

product E kill crawlinginsect
product E kill flyinginsect
product E kill hoppinginsect

product F kill annualweed
product F kill perennialweed
product F fertilize lawn

(this question has put more emphasis on intended use than the last, so the system has answered in more detail)

›› when should I use product F?

spring
fall

›› what can I use to kill snails?

product B
product C

›› what can I use to kill snails on my vegetables?

product C

(product B is not for use on food, and so is not suggested in this answer)

›› what can I use to kill weeds?

product Á
product F

(these are the suitable products known to the system)

›› what are the weeds?

annualweed
perennial weed

(the system has listed, without much detail, the kinds of weeds)

›› what are weeds?

sorry, I don't know

(we have reached a limit of the knowledge in the system)

›› what are the perennial weeds?

bahiagrass
bull thistle
clover
dandelion
kentucky bluegrass
poison ivy
poison oak
yellow oxalis
oxalis

(the system has understood this as a request for a detailed answer)

›› is there anything I can use that will fertilize my lawn?

product F

(a suitable product is suggested)

›› what can I use to kill weeds in my lawn in spring?

product F

(the same product can be used)

›› what can I use to kill weeds around my fence?

product A

(product F is for lawns, but elsewhere we can use something stronger)

›› do I need a sprayer to use product A?

yes

›› can I use product A without a sprayer?

sorry. no

›› what are the precautions I should follow in using product A?

avoid contact with eyes
keep children and pets away till spray has dried
avoid contamination of food
if accidental spray then wash off with water

›› what is the response time of weeds to product A?

annualweed 5 days
perennialweed 15 days

›› what is the response time of each product that kills annual weeds?

product A 5 days
product F unknown

›› what can I use on dandelions?

product A
product F

›› what can I use on dandelions that will kill them in less than 20 days?

product A

›› does product A kill dandelions in less than 20 days?

yes

›› does product A kill dandelions in less than 2 days?

sorry. no

This session shows that the Kbo1 system has considerable knowledge of the properties of a few products and their intended uses. As with all knowledge bases known to us, there are limits to the domain which is covered. However, when a question cannot be answered, the user sees reasonable replies such as "I'm sorry, I don't know", or "I don't understand the word: bus". Although the domain of competence is much smaller than that of most adult people, these phrases carry about the same information as an immediate reply from a person who does not know the answer to a question.

The Kbo1 program, which is written in PROLOG, consists of an English input component, a knowledge base of products and their intended uses, and an output component. The knowledge base and the output component are fairly specific to the task at hand, while parts of the input component have been written in a general way for re-use in other programs.

When a question is typed in, the input component first looks up each word in a dictionary (stored as a set of PROLOG facts), and finds the parts

of speech. Next, the question is parsed by a grammar consisting of a set
of PROLOG rules. As the parse proceeds, a representation of the meaning
of the question is constructed. If the parse is pursuing a 'garden path' which
cannot succeed, this is often detected early by the rules for meaning construction, causing the system to try a different parse.

The meaning of a question is represented as a PROLOG statement
either of the form 'is p(X) true in the knowledge base?' or of the form 'the
set of all X such that p(X) is true in the knowledge base', where X and p
may be structured terms. For example, the English question

    does product A kill dandelions in less than 20 days?

is translated by the input component into the PROLOG query

    yesno ! use(product A. kill-weed:dandelion. response(T)) & typedlt(T. 20.days)

which can be read as asking whether product A can be used to kill a weed,
in particular a dandelion, yielding a response time T, such that T is less
than 20 days.

Special care is taken to give informative, rather than yes–no, answers
to questions such as

    do you sell anything that kills bluegrass?

Very few people would be satisfied with only the answer 'yes', so the
input component generates the meaning

    set(item):I ! all(P.use(P.kill-weed:bluegrass. M). I)

which can be read as saying that what is required is the set I of products
P such that one can use P to kill a weed, in particular bluegrass, with no
time or other restrictions M. This retrieves from the knowledge base all of
the products which kill bluegrass.

The knowledge base component of Kbo1 contains an is-a hierarchy,
which is used to answer questions at an appropriate level of detail, and
some basic rules and facts about products. The hierarchy contains facts
such as

    weed:annualweed
    annualweed:bluegrass

that is, bluegrass is an annual weed, and an annual weed is a weed. The
information about products is stored in statements such as

    can _ use(product A. kill. weed:Y:Z) ‹-weed:Y:Z

which says that product A can be used to kill weeds, for instance bluegrass which is an annual weed, which is a weed.

The output component of Kbo1 contains information about which kinds of questions (yesno, set) require which kinds of answer format.

In constructing the Kbo1 system, it was not necessary to write an inference engine or a parser. Both of these items were covered by the use of the built in inference method of PROLOG. In terms of coverage of a part of English, the behavior of the system is promising. In terms of performance, we find that no question takes more than 1 second to answer, even when our computer is heavily loaded with work by other users. Our coverage and performance results are consistent with those reported (McCord, 1982; Warren & Pereira 1981) namely, that efficient PROLOG programs can be written for useful natural language access to knowledge bases. In summary, the direct representation of syntax, semantics and knowledge in the language PROLOG appears to be a good approach to the construction of useful knowledge-based systems which can answer questions in ordinary English.

5. Conclusions

This chapter has described three systems, written in in PROLOG, which span several areas of work in databases and expert systems.

The first system, SYLLOG, gradually accumulates knowledge of a task domain during normal use. While this knowledge acquisition is happening, SYLLOG checks incoming rules against the rules it has stored. SYLLOG automatically prompts with the sentences that it has been told about. Provided that certain simple conventions are followed, these sentences can be in English, or in any other language or specialized dialect which uses the same letters as English. Typically, a question is posed by selecting a sentence on the screen, and modifying it if necessary. A question can be about facts, or about the rules for using the facts. Both facts and rules can be modified on the screen. SYLLOG automatically provides explanations for its answers to yes–no questions.

The second system, PLANTDOC, is a MYCIN-style expert system for diagnosing problems with plants and for finding products or treatments to correct the problems. While SYLLOG reasons exactly, PLANTDOC reasons judgementally using confidence factors. A rule in PLANTDOC is similar to a rule in SYLLOG, but has an attached confidence factor and does not have example elements. The rules are placed in PLANTDOC off-line, rather than online during the use of the system. When the system is in use, it asks questions, generated from the rules, which can be answered with yes, unknown, no, or a confidence number. Before asking a question, the PLANT-

DOC inference engine checks whether the answer could affect the final advice it gives about a product or treatment. A question is only asked if it could make a difference.

The third system, Kbo1, answers questions in English about products in a garden store and their uses. Unlike the other two systems, the Kbo1 prompt contains no information about what is in the system. However, Kbo1 can answer questions such as "what do you know about?" to get the ball rolling. When the limits of the knowledge in the system are reached, it gives an answer such as "I'm sorry, I don't understand". The system has a dictionary, an English syntax and semantics, and a knowledge base of rules and facts about products, all written directly in PROLOG. There is no parser or inference engine, as the built-in inference method of PROLOG covers both of these functions. Each question is answered immediately (in under one second) after it is asked.

The three systems are for knowledge-based access to data, for giving expert advice about plant problems, and for answering English questions about garden products, respectively. Each system is a collection of PROLOG facts and rules, and hence is rule-based.

One aspect of writing rules, rather than conventional programs, is that rules may be combined by an inference engine to produce advice that, while correct, has not been anticipated. As has been persuasively argued by Michie (1982), it is very important that any program be able to produce a convincing explanation of its conclusions. Rule-based programs represent knowledge in a way that makes this feasible. In an extension of PROLOG (Walker, 1983), explanations, at a controllable level of detail, are provided for the results from any PROLOG program. These explanations are mapped into the English-like sentences of SYLLOG, can be displayed as a summary of which rules were used to give advice in PLANTDOC, or can explain why certain products are or are not applicable in Kbo1.

There are many useful ways of representing knowledge in expert systems. Popular representations include frames, semantic networks, MYCIN rules, causal networks, and logic. It has been pointed out by Moore (1982) that there is a rather simple problem, concerning three blocks on a table, for which the only useful representation in the above list is logic. Moreover, each of the above representations maps rather neatly into logic. For example, (Kowalski, 1979) shows how this mapping can be done for semantic networks.

Up till recently, it has not been clear how to take collective advantage of the large amount of significant progress made by each separate school of knowledge representation. The PROLOG language provides a rather powerful and practical means for integrating some apparently diverse techniques in databases and expert systems.

6. References

Aikins, J. S., Kunz, J. C., Shortliffe, E. H., & Fallat, R. J. *PUFF: An expert system for interpretation of pulmonary function data* (Tech. Rep. STAN–CS–82–931). Computer Science Department, Stanford University, 1982.

Clocksin, W. F., & Mellish, C. S. *Programming in PROLOG.* New York: Springer-Verlag, 1982.

Codd, E. F. Relational database: A practical foundation for productivity. *Communications of the ACM,* February 1982, *25*(2), 109–117.

Ennis, S. P. Expert systems, a user's perspective of some current tools. *Proceedings of the National Conference on Artificial Intelligence,* 1982, 319–321.

Forgy, C. L. OPS–5 *user's manual* (Tech, Rep. CMU–CS–81–135). Department of Computer Science, Carnegie-Mellon University, 1981.

IBM. *SQL/data system concepts and facilities* (Tech. Rep. GH24—5013). IBM Corporation, 1981.

Kowalski, R. *Logic for problem solving.* New York: Elsevier North-Holland, 1979.

McCord, M. Using slots and modifiers in logic grammars for natural language. *Artificial Intelligence,* 1982, *18,* 327–367.

Michie, D. Game playing programs and the conceptual interface. *ACM Sigart Newsletter* No. 80, 1982, 64–70.

Moore, R. C. The role of logic in knowledge representation and common sense reasoning. *Proceedings of the National Conference on Artificial Intelligence,* 1982, 428–433.

Pereira, L. M., Sabatier, P. & Oliveira, E. Orbi: An expert system for environmental resources evaluation through natural language. *Proceedings of the First International Logic Programming Conference,* 1982, 200–209.

Selinger, P. G., Astrahan, M. M., Chamberlin, D. D., Lorie, R. A., & Price, T. G. Access path selection in a relational data base management system. *Proceedings of the ACM SIGMOD,* 1979, 23–34.

Shortliffe, E. H. Computer-Based Medical Consultations: MYCIN. New York: Elsevier North-Holland, 1976.

Van Melle, W., Scott, A. C., Bennett, J. S., & Peairs, M. *The EMYCIN manual.* (Tech. Rep. STAN–CS–81–885). Computer Science Department, Stanford University, 1981.

Walker, A. *SYLLOG: A knowledge based data management system* (Tech. Rep. 034). Computer Science Department, New York University, 1981.

Walker, A. *PROLOG/EX1: An inference engine which explains both yes and no answers. Proceedings of the Eighth International Joint Conference on Artificial Intelligence,* 1983, 526–528.

Walker A. *SYLLOG: an approach to PROLOG for nonprogrammers.* In M. van Caneghem & D. H. D. Warren (Eds.), *Logic programming and its applications.* Norwood, NJ: Ablex, 1984.

Walker A., & Porto, A. *Kbo1: a knowledge based garden store assistant* (Tech. Rep.). San Jose, CA: IBM Research Laboratory 1983.

Warren D. H. D., & Pereira, F. C. N. *An efficient easily adaptable system for interpreting natural language queries* (DAI Research Paper No. 155). Department of Artificial Intelligence, University of Edinburgh, 1981.

Zloof, M. M. Query-by-Example: a data base language. *IBM Systems Journal,* 1977, *16*(4), 324–343.

AI and Decision Making: The PROSPECTOR Experience*

Richard O. Duda
René Reboh
SYNTELLIGENCE
800 OAK GROVE AVENUE
MENLO PARK, CALIFORNIA 94025

PROSPECTOR is an expert system that was designed for decision-making problems in mineral exploration. It uses a structure called an inference network to represent the required judgmental knowledge. Once encoded, this knowledge can be used in a variety of ways. This chapter describes the use of inference networks as a language for representing and using expert knowledge, and presents examples showing its use in three different kinds of decision-making problems.

1. Introduction

There are numerous signs that the efforts of many years of research in artificial intelligence (AI) are beginning to bear fruit. Established industrial corporations and new startup companies are seeking ways to apply AI methods of knowledge representation and problem solving to tasks ranging from industrial robotics to natural language understanding.

Some particularly effective performance has been achieved by a class of programs that have come to be known as *expert systems*. Like the phrase

* When the first version of this paper was prepared, R. O. Duda was affiliated with Fairchild Camera and Instrument Corporation and R. Reboh was affiliated with SRI International. The research that we describe was performed at SRI International, and was supported by a series of contracts with the U.S. Geological Survey and the National Science Foundation; any opinions, findings, conclusions, or recommendations expressed in this publication are those of the authors, and do not necessarily reflect the views of either the USGS or NSF. Many computer scientists and many consulting geologists contributed to PROSPECTOR, and the work we have described represents their efforts as well as ours. The contributions of Peter E. Hart, Kurt Konolige, John Reiter, and the late John Gaschnig are particularly relevant, and are greatly appreciated.

"decision support system," the phrase "expert system" has been applied to so many different programs that it is in danger of loosing its meaning. Broadly speaking, however, we can say that an expert system is a computer program that uses explicitly represented knowledge and computational inference procedures to solve problems normally requiring sigificant human expertise.[1]

Of the many expert systems that have been developed in the last decade, the majority of the successful programs were designed to play a role analogous to that of a human consultant. Not surprisingly, this work was done in areas of medicine, chemistry, and geology in which there is an established tradition for consultation. Human consultants in these fields are valuable because they are specialists who possess extensive knowledge about particular problem domains.

Expert consultation systems have focussed on problems in which a human expert's knowledge is largely factual in nature. Here the key to solving a problem lies more in knowing the relevant information than an ingeniously constructing a solution from logical principles. The human expert is distinguished by knowing all of the factors that are important, and by possessing judgment in combining diverse considerations to reach a decision. It follows that a corresponding expert system must have effective ways to represent and employ large amounts of different kinds of knowledge bearing on specialized problems.

In this paper we use the experience we gained in developing the PROSPECTOR consultation system (Hart, Duda, & Einaudi, 1979) to illustrate how AI methods can be applied to aid this kind of decision making. We focus on PROSPECTOR's use of *inference networks,* a method for representing judgmental knowledge that is related both to rule-based programming (Waterman & Hayes-Roth, 1978) and to more traditional methods of decision analysis (Raiffa, 1968). In particular, we shall view inference networks as providing a simple language that an expert can use to specify both the knowledge that is relevant to a class of decision problems and — to some extent — how that knowledge should be used.

2. Approaches to Decision Making

When thinking of problems suitable for computational solution, it is natural to concentrate on numerical problems. After all, electronic computers were originally developed to compute ballistics tables, became commercially

---

[1] A large number of diverse systems satisfy this definition, and in this paper we cite only a small portion of this work; for surveys of expert systems research, see Buchanan (1982), Nau (1983), or Duda and Shortliffe (1983).

successful through automating business computations, and are extensively used for scientific and engineering calculations. It is still widely thought that computers cannot be applied to problems unless the problems can be cast in a conventional mathematical form. Thus, given a new problem, one usually asks if it can be expressed in such languages as algebra or differential equations or statistics or optimization theory.

These classical approaches have many virtues. In particular, they bring clarity and objectivity, and permit exploitation of the computer's enormous proficiency at arithmetic. However, not all problems are well suited to numerical treatment. Many decision problems that involve experience and judgment have to be coerced into fitting the traditional decision theoretic frameworks. For example, most of the attempts to employ decision theory in medical diagnosis have failed to produce acceptable results (Shortliffe, Buchanan, & Feigenbaum, 1979). As attractive as these methods are in theory, they do not reflect the way that people approach these kinds of problems.

From the beginning, researchers who were interested in AI have concentrated on developing and understanding nonnumerical methods for problem solving. While AI programs use numerical methods when they are appropriate, their power comes from exploiting the fact that computers can also perform symbolic inference. AI researchers have also stressed the importance of knowledge representation. Where conventional programs embed knowledge implicitly as algorithms or procedures, AI programs expose this knowledge explicitly through separate symbolic data structures. The encoded knowledge is usually called the *knowledge base,* and the set of inference procedures that operate upon the knowledge base is often called the *knowledge-base interpreter* or the *inference engine.*

Several advantages accrue to this separation:

1. The same knowledge can be used for more than one purpose. For example, a given knowledge base can be used to solve a particular problem, to provide an explanation for the solution, or to support computer-aided instruction about the problem.
2. The power of the program can be extended either by expanding the knowledge base or by adding facilities to the interpreter. In particular, this allows a large system to be developed incrementally.
3. The problem-solving mechanisms and system facilities of the interpreter can be applied to similar problem domains by replacing the old knowledge base by a knowledge base for the new domain.

To see how such advantages can be realized in practice, we shall focus on PROSPECTOR, emphasizing the particular mechanisms that PROSPECTOR uses for representing and employing knowledge for decision making.

In the next section we describe in some detail the principles of the inference networks used to create PROSPECTOR's knowledge base. This is followed by a description of PROSPECTOR's interpreter and its three modes of operation. The flexibility of this approach to decision making is illustrated by describing its application to three different problems: mineral-deposit evaluation, drilling-site selection, and parameter selection for simulation. We conclude with a brief discussion of PROSPECTOR's tools for knowledge acquisition, and the use of inference networks as a knowledge-representation language.

3. Inference Networks

3.1. Certainty and Probability

PROSPECTOR's approach to decision making adopts many of the attitudes of Bayesian decision theory. In particular, it assumes that the job of a decision maker is to make a selection from a finite number of prespecified alternatives. In a medical diagnosis problem, the goal might be the identification of a disease or the selection of a therapy. In a business application, it might be the choice of one of a set of investment alternatives. In one of PROSPECTOR's applications, it is the classification of an exploration prospect in terms of known types of ore deposits.

The hypothesis that a particular alternative is correct is called a *top-level* hypothesis. Let $H_1, H_2, \cdots, H_n$ be the n top-level hyphotheses. These hypotheses are assumed to be simple propositions—statements that are either true or false. Unlike the conventional Bayesian case, they are not required to be either mutually exclusive or exhaustive. In any decision problem, some evidence may give positive support to a particular top-level hypothesis, some may be neutral, and some may be negative. The amount of potential evidence may be quite large, but we assume that it too is finite. Let $E_1, E_2, \cdots, E_m$ denote the m pieces of potentially acquirable evidence, which are also assumed to be simple propositions. The problem, then, is to acquire evidence so that one can establish a top-level hypothesis.

In general, the truth or falsity of a top-level hypothesis will never be known with complete certainty. As relevant evidence is acquired, the certainty of belief in the top-level hypotheses changes. It is natural to express this degree of belief through a numerical measure. Thus, Prospector associates a subjective probability value with every hypothesis to measure the degree to which it is believed to be true. As in Bayesian decision theory, it is assumed that in the absence of evidence any hypothesis H has an initial or *prior probability* $P(H)$. The observation of a piece of evidence E changes this to the *posterior probability* $P(H \mid E)$.

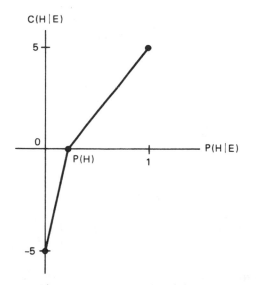

Figure 1. *Certainty and probability.*

PROSPECTOR also makes use of another certainty measure, a normalized probability that was introduced in the MYCIN system (Shortliffe, 1976; Shortliffe & Buchanan, 1975). This *certainty value C(H | E)* measures the degree to which acquired evidence *E* changes the probability of a hypothesis *H* from its initial or prior value *P(H)*. The definition is

$$
C(H \mid E) = \begin{cases} 5 \dfrac{P(H \mid E) - P(H)}{1 - P(H)} & \text{if } P(H \mid E) \geqslant P(H) \\[2ex] 5 \dfrac{P(H \mid E) - P(H)}{P(H)} & \text{if } P(H \mid E) < P(H). \end{cases}
$$

In words, $C(H \mid E)$ is a number between -5 and $+5$, where -5 corresponds to *H* being definitely false $(P(H \mid E) = 0)$ and $+5$ corresponds to *H* being definitely true $(P(H \mid E) = 1)$. The "no-information" case $(P(H \mid E) = P(H))$ always corresponds to a certainty of 0. Between these endpoints, there is a one-to-one linear relation between certainty and probability (see Figure 1).[2]

---

[2] Except for the factor of 5, this is the same as the certainty factor introduced by Shortliffe and Buchanan. We use a -5 to 5 rather than a -1 to 1 scale because the 11 natural levels of quantization allow convenient qualitative associations such as the following: (-5, certainly false) (-4, very probably false) (-3, probably false) (-2, unlikely) (-1, somewhat unlikely) (0, no opinion) (1, somewhat likely) (2, likely) (3, probably true) (4, very probably true) (5, certainly true). However, we have made no formal attempt to calibrate such interpretations.

In our experience, the main advantage of employing certainties instead of (or as well as) probabilities is psychological. For very good reasons, people often feel uncomfortable estimating prior probabilities. Yet, they are willing to say whether a piece of evidence increases or decreases the probability of a hypothesis with respect to its prior value, and are often willing to use the certainty value to estimate the amount of change. Thus, certainties are particularly useful as a technique for talking about relative probabilities. In addition, as we shall see later, they seem more natural than probabilities when establishing the context for a hypothesis.

3.2. Hierarchy and Network Structure

Let E' denote a collection of evidence that has been gathered at some point in time. To evaluate an hypothesis H, Bayesian decision theory calls for us to estimate the posterior probability $P(H \mid E')$. Unfortunately, two obstacles make this a far from straightforward procedure.

First, the available evidence is generally incomplete and uncertain. Consider, for example, the simple market forecasting hypothesis H: "During the next 12 months the price of oil stocks will rise." Obviously, many factors influence the odds on H, including the evidential assertion E: "OPEC is stable." We typically do not know and cannot afford to find out about every relevant factor, and we usually are uncertain about the very ones that are the most important.

Second, the probabilistic relations linking the hypotheses and the relevant evidence are both unknown and complex. Classical simplifying assumptions (such as conditional independence) are almost certainly wrong, and there is rarely enough data available to obtain good statistical estimates of the true relations.

Of course, some problems simply do not have solutions, and for these problems no methodology can succeed. However, there is hope when there are people—experts—who possess general knowledge that enables them repeatedly to make good decisions. When articulate experts are asked how they reach a decision, they often structure their explanations hierarchically. Even though a very large number of factors might bear on a decision, the expert will usually identify a small number of major considerations that more or less "independently" influence the decision. The determination of the state of these major factors is done through the same kind of breakdown into major subfactors, leading to a hierarchical decomposition of the decision procedure.

Inference networks support exactly this kind of hierarchical structuring. They do this in two ways. First, they provide a simple way to specify explicitly what the factors are and which factors affect which other factors.

Second, they provide a set of standard or primitive ways of computing the probability of a given factor from the probability of the factors that influence it.

The nodes in an inference network correspond to propositional assertions. The arcs incident on a given node N link N to the assertions that determine its probability; these assertions are called the *antecedents* for the *consequent* assertion N. In general, all assertions fall into three categories: top-level hypotheses, intermediate factors, or evidential statements. Intermediate factors serve as evidence for the nodes above them, and as hypotheses for the nodes below them. Thus, nodes with no outgoing arcs are top-level hypotheses, nodes with no incoming arcs are evidential statements, and nodes with both incoming and outgoing arcs are intermediate factors.

If there is only one path from any evidence node to any top-level hypothesis, then the network has a tree structure (Figure 2a). However, multiple paths are not unusual, in which case the inference network is a genuine graph (Figure 2b). However, to prevent "circular reasoning," the presence of loops is forbidden (Figure 2c). Thus, inference networks are acyclic graphs.

The prohibition of loops is the only absolute restriction on the topology of an inference network. However, in building large inference networks, we have found that there is good and bad structural style. For example, a very "bushy" and "flat" structure such as the one shown in Figure 2d is usually undesirable. This is the kind of structure one encounters in check lists, discriminant analysis, and similar procedures that attempt to combine all of the evidence at once; while all of the evidence should be considered, highly related pieces of evidence should be grouped together to limit the ways in which they can interact.

Generally speaking, whenever a node has, say, more than four of five antecedents, we have found it desirable to create new intermediate factors that separate the interactions of these antecedents. We have often found that these new factors correspond to well known concepts in the field. In the language of probability theory, this structuring is roughly analogous to approximating a complicated joint probability function by making an assumption of Markov dependence. However, the approximation is guided by subjective knowledge of the domain, rather than by strictly formal principles.

These considerations illustrate the major philosophical difference between the use of inference networks and traditional probability theory. In principle, probability theory is normative. While there may be many unknown parameters to estimate, once one has accepted the assumptions that define the probability structure, the values for the probabilities of the top-level hypotheses are implicitly determined. No considerations of taste

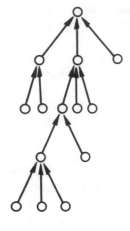

(a) A TREE

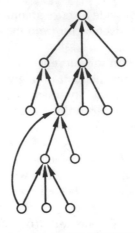

(b) AN ACYCLIC GRAPH

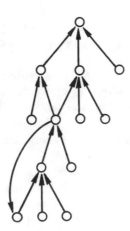

(c) A FORBIDDEN GRAPH

(d) AN UNDESIRABLE GRAPH

Figure 2. *Inference network topologies.*

or style arise. Whether or not any approximations used correspond well to reality, all of the axioms of probability theory will be honored.

By contrast, inference networks provide the expert with a language for more directly specifying how degrees of belief (expressed as subjective probabilities) are to be computed. The language does impose some constraints. As we shall elaborate in the next section, it requires that functions for computing probabilities be composed out of a small number of primitive functions. However, it also provides considerable freedom, such as allowing the specification of any loop-free network topology desired to group factors and control the flow of information.

The price paid for this freedom is that there is no longer any guarantee that all of the axioms of probability theory will be honored. However, if one views the values computed as heuristic measures of degree of belief, then the only question is whether or not it is easy to construct an inference network that adequately approximates the specifications of the expert.

3.3. Procedures for Combining Evidence

The topology of an inference network exhibits the paths through which information can flow in determining the probability of the top-level hypotheses. To compute an actual numerical value, we must also specify how the probability of any given node can be computed from the probabilities of its antecedent nodes.

PROSPECTOR provides two basically different mechanisms for combining the probabilities of the antecendents: (a) logical combination, and (b) weighted combinations. In addition, it provides a contextual mechanism that controls the application of these procedures. It is up to the designers of the inference network to decide which of these basic mechanisms to use, and how to combine them to compose an inference network that exhibits the desired behavior. To make those decisions, one must understand how the basic mechanisms behave, and thus we consider each of them in turn.

3.3.1. Logical combinations. It often happens that an assertion A can be expressed naturally as a logical or Boolean combination of its k component factors $A_1, A_2, \cdots, A_k$. The classical combinations are *conjunction* and *disjunction:*

$$Conjunction: A = A_1 \text{ and } A_2 \cdots \text{ and } A_k$$
$$Disjunction: A = A_1 \text{ or } A_2 \text{ or } \cdots \text{ or } A_k.$$

In a conjunctive combination, all of the component assertions must be true for their conjunction to be true; if any single component is false, the conjunction is false. In a dual fashion, in a disjunction all of the component assertions must be false for the disjunction to be false; if any single component is true, the disjunction is true. The operation of *negation* ($\bar{A}$ is true if and only if A is false) completes the logic, since it is well known that any Boolean expression can be expressed in terms of conjunction, disjunction and negation.

Typically, however, we do not know whether the components are true or false; all we have is a subjective estimate of the probability that they are true. Let E' denote the evidence that has been acquired at any given time, and let $P(A_i | E')$ be the current probability of A_i. It is easy to compute the probability of the negation of A_i, since

$$P(\bar{A}_i | E') = 1 - P(A_i | E').$$

Unfortunately, for conjunction and disjunction one needs to know the entire joint probability $P(A_1, A_2, \cdots, A_k | E')$ to compute $P(A | E')$ exactly, not just the marginal probabilities $P(A_i | E')$. The assumption of statistical independence leads to simple product formulas, but, in our experience, the results are usually too pessimistic for conjunction and too optimistic for disjunction.

This is typical of the dilemmas that confront attempts to apply probability theory rigorously to these kinds of problems. PROSPECTOR's "solution" is heuristic. The basic procedure is to use the following formulas from Zadeh's theory of fuzzy sets (Zadeh, 1965):

$$\textit{Conjunction:} \quad P(A | E') = \min_i \{P(A_i | E')\}$$

$$\textit{Disjunction:} \quad P(A | E') = \max_i \{P(A_i | E')\}.$$

Through the recursive application of these formulas and the equation for negation, the "probability" of any Boolean expression can be computed from the probabilities of its components.

These formulas have the characteristic that the least certain component dominates conjunction, and the most certain component dominates disjunction. In particular, if the probability of any component is still at its prior value, no amount of "good news" about the other components can increase the probability of a conjunction, and no amount of "bad news" can decrease the probability of a disjunction. If this behavior is not acceptable, then a more complicated inference network must be composed using either weighted combinations or contexts.

3.3.2. Weighted combinations. The use of weighted combinations is called for whenever it is appropriate to think of each of the k component factors, $A_1, A_2, \cdots, A_k$ as "casting a vote" to determine the degree to which an assertion A is true.[3] This is essentially what happens when statistically

---

[3] In PROSPECTOR, we often say that such weighted combinations correspond to *rules* of the form *IF* $A_1, A_2, \ldots A_k$ *THEN* (to some degree) A. Thus, PROSPECTOR is often described as a rule-based system. However, PROSPECTOR does not possess the full flexibility of a production system (Davis & King, 1976), since it contains only primitive mechanisms for binding variables in the assertions. In essence, computational generality has been sacrificed to obtain greater efficiency and greater control over the inference process.

ECindependent data are combined using Bayes' Rule. The formula is particularly simple when expressed in terms of odds and likelihood ratios instead of in terms of probabilities. Let $O(A)$ be the prior odds on A, where

$$O(A) = \frac{P(A)}{1 - P(A)},$$

and let λ_i be the likelihood ratio

$$\lambda_i = \frac{P(A_i \mid A)}{P(A_i \mid \bar{A})}.$$

Suppose further that the A_i are conditionally independent under both the hypotheses A and $\bar{A}$. Then it can be shown (Duda, Hart, & Nilsson, 1976) that Bayes' Rule states that the logarithm of the posterior odds $O(A \mid A_1, A_2, \cdots, A_k)$ is given by

$$\log O (A \mid A_1, A_2, \cdots, A_k) = \log O (A) + \sum_{i=1}^{k} \log \lambda_i .$$

Thus, we can think of each of the factors A_i as casting $\log \lambda_i$ "votes" for the log-odds of A. More precisely, the votes are in favor of A if the likelihood ratio is greater than 1, and against A if it is less than 1; if $\lambda_i = 1$, then the truth or falsity of A_i has no effect on the odds on A.

This form of Bayes' Rule assumes that each of the k components A_i is known to be true. If nothing is known about a particular factor, its log-likelihood can just be removed from the sum. However, what should be done if we only have some probability $P(A_i \mid E')$ that A_i is true, including the possibility that $P(A_i \mid E') = 0$, i.e., that A_i is false? PROSPECTOR's approach involves replacing each likelihood ratio by an *effective likelihood ratio* λ'_i. Like the certainty measure, the effective likelihood ratio is determined by three fixed points:

$$\hat{\lambda}_i = \begin{cases} \lambda_i & \text{if } P(A_i \mid E') = 1 \\ 1 & \text{if } P(A_i \mid E') = P(A_i) \\ \bar{\lambda}_i & \text{if } P(A_i \mid E') = 0 , \end{cases}$$

where $\bar{\lambda}_i$ is the likelihood ratio when A_i is known to be false:

$$\bar{\lambda}_i = \frac{P(\bar{A}_i \mid A)}{P(\bar{A}_i \mid \bar{A})} = \frac{1 - P(A_i \mid A)}{1 - P(A_i \mid \bar{A})} .$$

At intermediate points, λ'_i is determined by an interpolation procedure that is described in (Duda, Hart, & Nilsson, 1976).

In words, with weighted combinations there are two weights for every component factor. One weight (λ_i) applies when the assertion A_i is known to be true. The other ($\bar{\lambda}_i$) applies when A_i is known to be false. Between these extremes an intermediate weight is used, with A_i having no effect on the probability of A when nothing is known about A_i.

As the name implies, the use of weighted combinations allows different component factors to have different degrees of importance in determining belief in the conclusion. In choosing these weights one typically works with the factors one at a time, asking the expert to estimate $P(A \mid A_i)$ and $P(A \mid \bar{A}_i)$ for each factor independently, and computing the likelihood ratios from

$$\lambda_i = \frac{O(A \mid A_i)}{O(A)} \, ,$$

and

$$\bar{\lambda}_i = \frac{O(A \mid \bar{A}_i)}{O(A)} \, .$$

It might seem that logical combinations are merely a special case of weighted combinations. For example, if all of the ($\bar{\lambda}_i$ were zero, then $P(A \mid A_1, A_2, \cdots, A_k)$ would be zero if any of the factors A_i were false, and conjunction-like behavior would be obtained. However, there is a major difference. With weighted combinations the probability of the hypothesis can become quite high even though nothing is known about one or more of the component factors, while with conjunction it can never become higher than the unknown factor that is least likely *a priori*. Which of these behaviors is "right" depends upon the desired behavior, and this kind of reasoning is frequently used in deciding whether to use logical combinations, weighted combinations, or a mixture of the two.

3.3.3. Contexts. So far we have tacitly assumed that it makes sense to talk about any proposition independently of the state of any other proposition. However, it frequently happens that one proposition refers to another, and one cannot always meaningfully talk about the former in isolation. For example, one usually cannot consider the property of some object before the existence of that object has been established. Even when isolated reference is meaningful, it is often the case that an expert cannot provide strong rules unless the existence of some special situation has been established.

To handle such situations, PROSPECTOR allows any proposition C to be designated as a *context*. In the inference network, special arcs called context arcs link any dependent propositions A to C. A context arc blocks the upward propagation of any information about A if the proper context has not been established. Thus, if an important conclusion depends upon A, the inference network interpreter will set up the subgoal of first establishing the context C. To prevent the status of a context from switching from "established" back to "unestablished" during the course of subsequent information gathering, we require that all applicable methods for determining the probability of a context be attempted before it can be declared to be "established." At that point investigation of A can proceed; otherwise, this line of inquiry is blocked.

There is no general answer to the question of how high the probability of C must be to treat it as being established. One clearly should not use a uniform threshold, since some propositions have a very high prior probability. Our approach has been to let the expert set a threshold on the certainty value rather than the probability value. We have also found it convenient to generalize this notion, allowing the expert to specify any desired certainty interval in which the context condition is considered to be satisfied. This allows the enabling (or disabling) of parts of the inference network when a contextual situation is known to be present, absent, or unknown.[4]

It might be thought that something essentially equivalent to this context mechanism could be constructed out of logical and/or weighted combinations. For example, one might merely include C in a conjunction with every A that has C as a context. However, there are two differences in behavior. First, the probability of the answer would then depend upon both the probability of C and the probability of A, whereas with contexts the probability of C has no effect on the answer once the contextual condition is satisfied. Second, and more important, there is no guarantee that the system would attempt to establish the context C before considering the dependent factor A.

At this point it becomes clear that the context mechanism goes beyond the representation of factual knowledge into the area of *control* — the specification of the sequence in which that knowledge is to be employed. For example, the expert can (and sometimes does) specify a certainty interval of -5 to 5 merely to force the consideration of C before the consideration of A. Thus, inference networks are more than just a declarative data structure, but contain some of the elements of procedural programming

---

[4] MYCIN employs a similar mechanism through its use of predicates such as KNOWN, NOT-KNOWN, DEFINITE, MIGHTBE, THOUGHTNOT, and so on (Shortliffe, 1976); traditional probability theory does not seem include any comparable formalism.

languages as well. It is sometimes tempting to abuse these facilities and to start composing programs in the knowledge base, thereby losing the advantages of separating the knowledge base from its interpreter. Resolution of these questions is one more example of the issue of proper programming style.

4. The Inference Network Interpreter

4.1. Modes of Operation

As we have seen, an inference network represents knowledge about a decision problem by explicitly exhibiting (a) the important propositional assertions, (b) the paths for propagating information from assertion to assertion, and (c) the procedures to be used for updating probabilistic measures of belief in the assertions. However, as it stands, such a knowledge base is essentially a static structure. What brings it to life is the network interpreter.

Some core parts of the interpreter are the same for any application. For example, if the probability of any assertion is changed, then the procedure for appropriately propagating that change upward through the arcs is common to all uses. However, other parts are specialized for particular applications.

PROSPECTOR has three basic modes of use: (a) interactive consultation, (b) batch processing, and (c) compiled execution. In the interactive mode, the program must choose an effective line of questioning, provide explanations for decisions, and respond to user commands. Batch mode is used for off-line processing of questionnaire data, or for systematic validation or sensitivity studies. Compiled mode is used when efficiency of execution is essential. We discuss each of these cases in turn.

4.2. Interactive Consultation

It is traditional to think of expert systems as serving as interactive consultants—programs that try to play a role analogous to that of a human consultant who is accessible only by telephone (or, for a more realistic metaphor, only through a computer-terminal link). During interactive consultation, the program itself is operating in one of two modes — the *antecedent mode* or the *consequent mode*. In the antecedent mode the program is receiving information; it matches statements from the user against the assertions in the inference network and propagates probability changes up through the network. In the consequent mode the program is attempting to establish (or rule out) a top-level hypothesis, and it searches the inference network for evidence nodes that are most effective for this task. The strategy that the program uses to choose goals and select evidence is called the *control strategy,* and we consider each mode of the control strategy in turn.

4.2.1. The Antecedent Mode. A typical consultation session begins with the user volunteering information in the form of simple assertions. A full exposition of the methods PROSPECTOR uses to represent and match assertions is beyond the scope of this paper.[5] Suffice it to say that given an acceptable users' assertion U and a description of a node N in the inference network, a semantic network matcher is able to determine whether

1. U and N are equal,
2. U is a subset of (describes a special case of) N,
3. U is a superset of (describes a general case of) N, or
4. None of the above relations holds.

Case 1 is the simplest; when the user volunteers the truth of an assertion in the inference network, the probability of that assertion can be set to 1 and the consequences can be propagated.[6] In Case 2 the user has said more than the inference network requires; if the user's assertion was positive, i.e., if the probability of U is greater than the probability of N, then the system can again propagate probabilities. Unfortunately, both of these cases are rather rare. Case 3 is more common; unless the assertion was negative, the system needs more information before it is logically permitted to propagate probabilities. Nevertheless, the fact that the user said something that was relevant should not be ignored. Case 4 is also common, and usually amounts to a statement whose relevance is unclear; these statements are ignored.

Thus, a typical consequence of initial volunteering is that a few exact and a number of partial matches are made between volunteered evidence and the nodes in the inference network, and some changes occur in the

---

[5] PROSPECTOR's use of semantic networks to represent assertions is described in Duda, Hart, and Nilsson (1978). The semantic network matcher is described in Reboh (1980); for other uses of the matcher in PROSPECTOR, see Reboh (1981).

[6] Even this case is not quite so simple, since the user can associate a certainty with his or her assertion, and the certainty can be negative as well as positive. Furthermore, if the node that was matched is an intermediate factor, subsequent evidence might well change this certainty. To accommodate these possibilities, PROSPECTOR does not merely reset a probability, but instead generates a new weighted arc, with the weight being computed so that the present certainty agrees with the user's certainty. This leads us into the thorny question of how contradictions in the user's inputs should be treated. PROSPECTOR has an ability to detect direct contradictions, and will insist that the user modify his or her statements to make them consistent; more subtle contradictions can pass undetected, however.

probabilities of the top-level hypotheses. At this point PROSPECTOR turns to the problem of working interactively with the user to choose a top-level hypothesis for further refinement. For each top-level hypothesis a score is computed that combines the certainty of that hypothesis and an average "effective" certainty of related nodes that were partially matched by volunteered evidence.[7] The top-scoring hypotheses are revealed to the user, who is then given the choice of either following the program's recommendation or selecting a different hypothesis to pursue. In either case, the system has a goal hypothesis to work on.

This is certainly not the only rational way to proceed with the investigation. In particular, when there are many top-level hypotheses, it is clearly desirable to group them so that a "divide-and-conquer" procedure can dispose of a number of hypotheses at a time. However, this alternative is not effective when the top-level hypotheses are not mutually exclusive. The INTERNIST–1 system has paid particular attention to the problem of setting up differential diagnosis tasks in problems where several top-level hypotheses can simultaneously be true (Pople, 1982). By enlisting the user's help on selecting hypotheses, we have adopted a rather different approach — one that assumes that the user wants to play and is capable of playing a role in the decision process. This also helps him to think of PROSPECTOR as more of a tool than an "expert in a black box."

4.2.2. The Consequent Mode. Once a top-level hypothesis H has been chosen, the program enters the consequent mode. Here it searches the inference network below H to determine what question to ask the user to help resolve the issue. In so doing, PROSPECTOR attempts to satisfy the following criteria:

1. *Effectiveness.* The evidence sought should have the potential to make a difference in the conclusion.
2. *Naturalness.* The program should not jump from topic to topic in a disorganized way that might confuse the user, but should follow a coherent line of reasoning.
3. *Responsiveness.* Interactive consultations should not be delayed by time-consuming optimization calculations.

PROSPECTOR's strategy for satisfying these requirements is based on backward chaining. Fundamentally, it recursively looks for the antecedent node that has the greatest effect on the current consequent node, stopping when it reaches a new node that it can ask about. However, there are some details that add interest to the procedure.

To begin with, contextual conditions must be honored. Given a goal node N (which is initially the hypothesis H selected by the user), the program first checks to see if N has any unsatisfied contexts; if so, establishing the

---

[7] The exact scoring function is given in Appendix C of Duda et al., (1979).

first such context becomes the new goal. When a goal node N with satisfied contextual conditions is found, the type of the node is considered. In general, any node will fall into one of three categories:

1. Terminal nodes that represent basic data that the user should, in principle, be able to supply.
2. Nonterminal nodes that correspond to logical combinations of other nodes. These nodes are almost always unaskable.
3. Nonterminal nodes that correspond to weighted combinations of other nodes. The expert who built the inference network can declare these nodes to be askable or unaskable.

If N is a terminal node, the system asks the user about N and propagates the results. If N is a logical node, the system selects the first "unexhausted" antecendent of N as the new subgoal and the procedure is repeated.[8] If N is a weighted-combination node, is askable, and has not been asked about before, the system asks the user; otherwise the best antecedent of N becomes the new subgoal.

The only remaining question is how one defines the "best" antecedent of an hypothesis N. Several considerations are clearly important: the values of the likelihood ratios λ and $\bar{\lambda}$, the current effective likelihood ratio λ', the current probability $P(E \mid E')$ that the antecedent evidence E is true, and the degree to which the hypothesis H is currently believed or disbelieved. For example, if the current hypothesis is very strongly suspected but there is an antecedent E having a very small $\bar{\lambda}$, so that belief in H might be substantially weakened if E were not true, then E is an interesting candidate — but not if it is very likely that E is true. After some experimentation, we found that the following criterion function provides an effective way to score the antecedents:

Case 1: $\lambda > \bar{\lambda}$

$$J^* = \log\frac{\lambda}{\lambda'} * P(E \mid E') * [1 - MB(H \mid E')]$$

$$+ \log\frac{\lambda'}{\lambda} * [1 - P(E \mid E')][1 - MD(H \mid E')]$$

Case 2: $\lambda \leqslant \bar{\lambda}$

$$J^* = \log\frac{\bar{\lambda}}{\lambda'} * P(E \mid E') * [1 - MD(H \mid E')]$$

$$+ \log\frac{\lambda'}{\lambda} * [1 - P(E \mid E')][1 - MB(H \mid E')]$$

---

[8] A node is "exhausted" if (a) it has no descendants and it has been asked about before, or (b) all of its descendants are exhausted.

where *MB* and *MD* are the measures of belief and disbelief defined by Short-liffe and Buchanan (1975) as

$$
MB(H \mid E') = \begin{cases} \dfrac{P(H \mid E') - P(H)}{1 - P(H)} & \text{if } P(H \mid E') > P(H) \\[2em] 0 & \text{if } P(H \mid E') \leqslant P(H) \end{cases}
$$

and

$$
MD(H \mid E') = \begin{cases} \dfrac{P(H) - P(H \mid E')}{P(H)} & \text{if } P(H \mid E') < P(H) \\[2em] 0 & \text{if } P(H \mid E') \geqslant P(H) \end{cases}
$$

Initially all probabilities are at their prior values, so that $\lambda' = 1$ and $MB = MD = 0$. In that state, J^* favors antecedents with small likelihood ratios (necessary conditions) if the evidence is unlikely to be found, and antecedents with large likelihood ratios (sufficient conditions) if the evidence is likely to be found. As the measure of belief MB increases, antecedents that reduce belief are favored; conversely, as MD increases, antecedents that increase belief are favored. Although simple, this has proved to be a very effective criterion for antecedent selection.

4.3. Batch Processing

While interactive operation presents the most interesting opportunities for the network interpreter to display intelligent behavior, it must also support important daily, routine applications that are off-line. For example, Gaschnig (1982) reported the results of validation experiments and a pilot study project that involved repeatedly processing data for many more cases than it would have been possible to treat interactively.

To support this mode of use, PROSPECTOR includes several system facilities that we shall only mention briefly. First, to allow users who do not have convenient access to terminals to provide data, PROSPECTOR can automatically generate questionnaires for any inference network. When the data entered in a questionnaire are later transferred to a file, the program can read the answers from the file as if it were in normal interactive mode. Moreover, once the data are recorded, there are facilities for performing systematic studies, such as determining the sensitivity of the conclusions to various kinds of changes in the inputs or the likelihood ratios.

One of the more interesting of the comparisons reported by Gaschnig concerned the use of the system to analyze differences of opinions of different users about a particular case. In that example, two users independently answered the more than 100 questions on a questionnaire. Considering their many differences of opinion, it is not surprising that the program arrived at two very different assessments for the same case. However, subsequent sensitivity analysis revealed that only a few key questions were really important, and these could be resolved without becoming distracted by the second-order disagreements. The ability to perform such systematic comparisons will probably turn out to be one of the most valuable features of expert systems when they enter routine use.

4.4. Compiled Execution

There are several situations in which one knows beforehand which questions a "user" will be able to answer without knowing the exact answers. One instance arises in the processing of questionnaire data. Another occurs in the screening of databases. While one can certainly read answers from a file, all of the time spent by the processor in the consequent mode trying to devise an intelligent questioning sequence (and much of the time spent propagating the consequences of the answer to each question throughout the inference network) is wasted. While this inefficiency may not be consequential if only a few tens or even a few hundreds of cases are to be processed, it becomes serious when tens of thousands of cases are encountered, as in the map-processing application described in section 5.2.

To solve this problem, Konolige implemented an inference network compiler for PROSPECTOR (Konolige, 1979). The compiler assumes that there is only one top-level hypothesis,[9] and takes advantage of the fact that inference networks are acyclic. Under these conditions, the topology of the inference network defines a partial ordering of the nodes from the terminal nodes up to the hypothesis node. This partial ordering specifies that the probabilities of certain nodes must be computed before the probability of subsequent nodes can be determined. The compiler generates a particular linear ordering from this partial ordering, specifying an exact sequence in which node probabilities should be computed. It also generates efficient assembly-language code for executing these computations. For large inference networks, the resulting compiled code typically runs four orders of magnitude faster than the interpreted code.

---

[9] In the map processing application, there is only one top-level hypothesis whose probability must be computed repeatedly. The procedure can certainly be extended to a multiple-hypothesis case.

5. Applications

As we have seen, inference networks provide a general framework for representing knowledge for decision making, and the inference network interpreter provides the means for putting that knowledge to work. However, our discussion up to this point has been quite abstract. There is an art to employing these methods effectively, and that aspect of PROSPECTOR is best conveyed through specific examples. Thus, in the remainder of this paper we examine three different applications of this methodology to give concrete illustrations of its versatility.

PROSPECTOR was originally developed to aid in problems of mineral exploration, particularly in the area of regional resource assessment (Hart, Duda, & Einaudi, 1979). It has subsequently been applied to several other problems, including the three problems described in this section: prospect evaluation, drilling-site selection, and numerical simulation. Each of these applications illustrates a different one of PROSPECTOR's capabilities. Prospect evaluation makes use of the interactive consultation mode, drilling-site selection exploits the inference network compiler, and numerical simulation employs an extension of inference network techniques to treat propositions about numerical quantities.

Since it is hard to understand these applications without some understanding of the domain, in each case we begin by providing some background information.

5.1. Interactive Consultation for Prospect Evaluation

5.1.1. Background on Ore-Deposit Models. Exploration geology provides a classical example of decision making under uncertainty. Centuries of mineral exploration have led to the discovery of the obvious ore deposits, such as the ones where ore minerals outcrop at the surface. Thus, geologists in search of new ore bodies must look for the more subtle clues that might indicate the presence of a deposit at depth. The odds are not good. A commonly stated estimate is that a thousand prospects must be examined before one is found that can be developed into a profitable mine. Expressed in more personal terms, this means that a typical exploration geologist never finds a mineable deposit in his or her professional life. From the point of view of decision support aids, this means that error-free performance is not required; aids that merely improve the odds of discovery can be quite valuable.

In recent years, geophysics and geochemistry have provided measurements that provide many clues to the possible presence of ore bodies at depth. Such measurements supplement the standard geological evidence that comes from mapping rock structures, ages and compositions. However, the proper interpretation of all of this evidence requires experience and an understanding of the processes of ore-deposit formation. Years of study of ore deposits have led to their classification into various families, and professional careers are devoted to detailed studies of some particular class of ore deposits. While the different classes grade into one another, and while expert geologists have different opinions on how deposits should be grouped into families, almost all of the world's economically significant deposits fall in one of some 100 to 150 different categories. Twelve of the major categories and some of their subcategories are listed in Table 1.

In exploration geology, an ideal instance of any of these kinds of deposits is called an *ore-deposit model.* A model can be genetic or descriptive; the former embodies a theory of ore-deposit formation, while the latter merely summarizes empirical observations. In either case, a central problem is to decide the degree to which data on a particular prospect seem to match one of these models. If relatively inexpensive surface observations indicate a good match with an economically important deposit type, interest in the prospect is greatly increased; a sufficiently good match may lead to the second and much more expensive phase of exploratory drilling. On the other hand, a poor match — for reasons other than lack of important data

Table 1
Major Categories of Mineral Deposits

Volcanogenic Massive Sulfides *Archean *Kuroko *Beshi *Cyprus	Copper-Zinc Shales *White Pine *Belt Cu-Ag
	Tungsten-Bearing Skarns
Mississippi-Valley Lead-Zinc *Pine Point *Midcontinent	Tin Granites and Skarns
	Platinum-Group Metals
Porphyry Copper *Continental margin—Bingham *Continental margin—Cerro de Pasco *Island arc *Skarns	Chrominum *Stratiform *Podiform
Porphyry Molybdenum *Rhyolite type *Quartz-monzonite type	Precious Metals *Carlin-type low-grade Au *Homestake-type Precambrian Au *Au-qtz veins in metamorphic rock *Disseminated low-grade Ag
Nickel Sulfides *Layered intrusions *Komatiitic *Rift related *Plutonic	Uranium *Roll front *Grants *Unconformity-related vein type *Caldera

— may well divert attention to other prospects. Thus, the central problem for PROSPECTOR is to encode expert knowledge about ore-deposit models in a form that allows this degree of match to be determined.

For illustrative purposes, we shall give a highly simplified description of a typical ore-deposit model — the Yerington-type (or Type-A) porphyry-copper model described in more detail in (Duda, Hart, Nilsson, Reboh, Slocum & Sutherland, 1977). A contemporary theory holds that such deposits were formed as a consequence of the subduction of an oceanic plate at the margin of a continent. The resulting heat and pressure led to the formation of magma that worked its way upward. If zones of weakness extended to the surface, the magma might erupt through volcanoes, such as those along the west coast of North and South America. Intrusive systems that did not reach the surface would cool slowly, forming large-grained crystals and various characteristic zones of altered rock. In favorable regimes of pressure and temperature, ore minerals would crystallize in characteristic sequences. Unfortunately, this ideal pattern is usually fragmented by such subsequent processes as uplift, faulting, erosion, leaching, and burial. Thus, one must read through such transformations to reassemble the pieces of the puzzle and recognize a possible deposit. This requires additional expert judgment that one would like to capture in a computer program.

5.1.2. Encoding an Ore-Deposit Model. Inference networks provide a natural mechanism for representing much of this geological knowledge in a form suited for decision making. In our example, the top-level hypothesis H is that the data match the model for a Type-A porphyry-copper deposit. The probability $P(H \mid E')$ is interpreted as the degree to which the available data E' support this hypothesis.

The first problem in creating the inference network is to identify the assertions A_i to be considered in determining the final evaluation. In this example, our expert geologist judged that there were three major considerations:

1. A_1—the favorability of the petrotectonic setting,
2. A_2—the favorability of the regional environment, and
3. A_3—the favorability of the intrusive system.

Each of these primary factors summarizes the conclusions of a large amount of evidence, and the problem of determining their degree of truth is similar to the original decision problem. Thus, secondary factors that support these primary factors must be identified. For example, the favorability of the petrotectonic setting is determined by (A_{11}) the proximity of the prospect to a continental-margin mobile belt, (A_{12}) the presence of magmatic activity related to subduction, and (A_{13}) the age of the belt. This

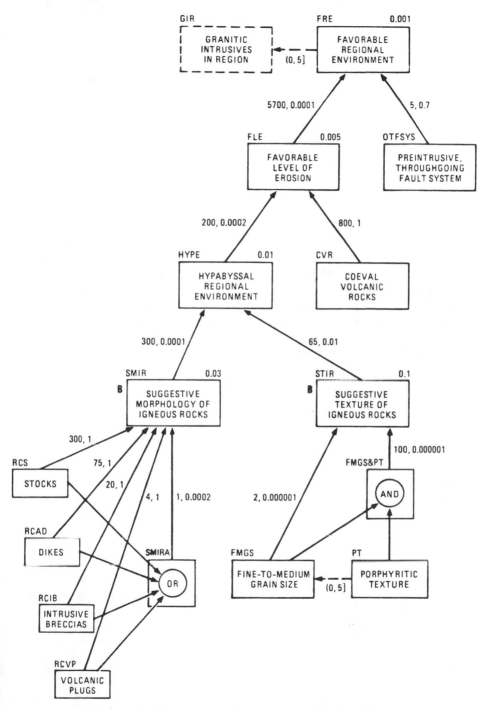

Figure 3. Inference network for part of an ore-deposit model.

process of recursive refinement continues until the supporting factors correspond to what should be directly observable field evidence.

A diagram of the actual inference network used to determine the favorability of the regional environment is shown in Figure 3. For our purposes, the geological interpretation of this diagram can be ignored. The main point to note is how Prospector's mechanisms for combining evidence have been employed.

In this diagram, conjunction and disjunction are indicated by the *AND* and *OR* nodes, weighted combinations are indicated by the solid arcs bearing pairs of numbers (λ and $\bar{\lambda}$), and contexts are indicated by the dotted arcs bearing certainty intervals. The four-node subnetwork at the lower right illustrates a typical inference network construction. This subnetwork expresses the expert's desire to see a certain kind of size distribution for the rock crystals known as "porphyry texture." The average grain size is weakly suggestive, and the mere observation of rocks with "fine-to-medium grain size" doubles the odds ($\lambda = 2$). In addition, it establishes the context for determining whether or not the rocks have the desired porphyritic texture. The combination of both of these factors multiplies the odds by 2 * 100, whereas the absence of either factor reduces the odds almost to zero ($\bar{\lambda} = .000001$). This composition of logic, weighted combinations, and contexts was crafted to combine these two pieces of evidence in accordance with the expert's assessment of how they should be treated, and is typical of the way that inference networks are constructed.

5.1.3. Interactive Consultation. Once the inference network is encoded, it can be used in any of the modes that we described in section 4. For the application to prospect evaluation, the interactive mode is natural. Here we can imagine the typical user to be an exploration geologist who has just spent several days making a surface examination of a prospect, who has found evidence that sparks his or her interest, and who would like to receive advice from a specialist in evaluating the degree to which the prospect matches a classical model.

The program begins by allowing the user to enter information about the significant types of rocks and minerals that have been observed or suspected to be present. This antecedent mode helps in the selection of models to pursue, which are scored and initially ranked. Once the user selects a particular model, the program enters into the consequent mode and begins asking a series of questions. The following transcript shows the initial questioning when the Type-A porphyry-copper model is selected.

The following questions are intended to establish the nature of the petrotectonic setting. (These questions will not be pursued unless a positive answer is given to the next question.)

1—To what degree do you believe that the target area is in a continental-margin mobile belt? *4*

2—Do you know the age of the continental-margin mobile belt? *WHY*
The vast majority of porphyry copper deposits are younger than 200 million years. Major age groups include:

   1. An older group of 200 to 150 m.y. (million year) porphyries in British Columbia;
   2. An important younger group of 70 to 50 m.y. deposits in the Mid-East, British Columbia (Skeena Arch), and Arizona-New Mexico Sonora (Laramide) copper province.
   3. The major belt of 60 to 4 m.y. deposits in the Andes of Argentina, Chile, Peru, Ecuador, and Columbia.

Although some major deposits in the western U.S. do not fit the above groupings, either in terms of age or location, they are all younger than 250 m.y. Known porphyry deposits of Paleozoic age (e.g., in the Appalachians and East Australia) constitute a minor fraction of present resources. Hercynian (late Carboniferous) occurrences in the Kazakhstan, USSR, may include some major deposits.

2.—Do you know the age of the continental-margin mobile belt? *YES*
3.—What is the age of the continental-margin mobile belt? *CRETACEOUS*
= 136 − 65 million years. What is your confidence in this estimate? *4*

This excerpt illustrates the nature of the interaction. Most of PROSPECTOR's questions expect Yes/No or certainty answers, although some questions ask for quantities; in the latter case, the likelihood ratio for the rule is a function of that quantity. If the user does not know the answer to a question, a certainty of zero leaves the probability at its prior value; although this weakens the certainty of the final conclusion (and, by failing to establish contexts, may prevent investigation of important issues), it does not prevent the program from proceeding.

In the consequent mode, PROSPECTOR, like MYCIN, allows the user to execute commands as well as to answer questions. Thus, in resonse to the WHY command at Question 2, the program accessed some explanatory text that the expert had previously prepared to explain why a particular piece of evidence is important. Other commands allow the user to do such things as trace internal inferences, change previous answers, change top-level goals, and obtain summaries of conclusions reached up to that point. For example, the following conclusions were eventually obtained for this particular example run:

I suspect that (* there is a Type-A porphyry copper deposit) (2.65)
There are several favorable factors; in order of importance:
   1. the petrotectonic setting is favorable for a type-A porphyry copper deposit 3.626
   2. there is a favorable regional environment for a type-A porphyry copper deposit 1.866
   3. there is a favorable intrusive system for a type-A porphyry copper deposit 2.607
   dominating factor
There are two positive factors with neutral effect that, if negative, could have been significant; in order of importance:
   4. You were sure that there are granitic, calc-alkaline intrusives in the region 5.0
   5. You were sure that the target (or prospect) lies in or near an intrusive system 5.0
For which of the above do you wish to see additional information?

Thus, the summary gives a top-down view of the considerations that go into the final evaluation. The user can explore any conclusion further by systematically stepping through the inference network. In general, the certainty of a conclusion will change if the certainty of any of its supporting factors changes, and the factor that has the greatest effect on the conclusion is identified as the "dominating factor." Although no systematic methods for sensitivity analysis are provided in the consultation mode, "what-if" experiments can be performed by changing answers and seeing how the changes affect the conclusions.

5.1.4. Status. PROSPECTOR's coverage of ore deposit models is incomplete. Inference networks have been created for nine prospect scale models like the one described above. Table 2 shows that a typical model contains about 150 nodes and 100 rules.[10] In addition, Table 3 lists 23 somewhat smaller regional-scale models that have been developed for regional resource assessment studies.

When the research effort first began, numerous false starts were encountered in learning the art of encoding models in inference networks. Thus, the first models tended to be small, frequently revised, and time consuming to develop. Unexpectedly large amounts of time had to be devoted to testing the models to be sure that they behaved as desired, and none of the models has been tested as thoroughly as we would wish.

Table 2
Statistics for the Prospect-Scale Models—June, 1983

Name	Description	Nodes	Rules
MSD	Massive Sulfide Deposit, Kuroko Type	39	34
MVTD	Missippi-Valley-Type Lead-Zinc	28	20
PCDA	Near-Continental-Margin Porphyry Copper, Yerrington type	186	104
PCDB	Near-Continental-Margin Porphyry Copper, Cerro de Pasco Type	200	157
PCDC	Near-Continental-Margin Porphyry Copper, Island-Arc Type	159	116
KNSD	Komatiitic-Type Nickel Sulfide	127	72
WSSU	Western States Sandstone Uranium	200	148
ECSU	Epigenic Carbonaceous Sandstone Uranium	197	153
LCU	Lacustrine Carbonaceous Uranium	164	111
PCDSS	Porphyry Copper Drilling Site Selection	133	60
VCPMDSS	Porphyry Molybdenum Drilling Site Selection, Vertical-Cylinder Type	57	42
HPMDSS	Porphyry Molybdenum Drilling Site Selection, Hood Type	76	48
Totals:	12 Prospect-Scale Models	1566	1065

---

[10] The last three models listed in Table 2 are for the drilling-site selection application described in the next section.

Table 3
Statistics for the Regional-Scale Models—June, 1983

Name	Description	Nodes	Rules
TCHD	Carbonate-Hosted Epigenetic (Cu)Pb-Zn(Ag), Tintic Type	71	57
MCHD	Carbonated-Hosted Epigenetic (Cu)Pb-Zn(Ag), Manto-Type	70	56
CSPBZN	Carbonate-Hosted Stratabound Lead-Zinc	57	37
SAVCHD	Southern Appalachian Valley Carbonate-Hosted Zinc	79	78
RCD	Dolomite-Hosted Copper (Ruby Creek)	53	39
RFU	Roll-Front Uranium	185	169
RWSSU	Western-States Sandstone Uranium	205	152
SPB	Sandstone-Hosted Lead	35	32
RMBSD	Restricted Marine Basin Shale-Hosted Cu-Pb-Zn (Ag,U,V)	38	26
SMRSD	Shallow Marine Rift Shale-Hosted	47	39
LSED	Lacustrine or Cutoff Marine Basin Shale-Hosted Evaporate (+ oil shale)	35	34
GT	Graywacke/Turbidite Zn-Cu-Pb(Hg, Fe, S)	53	31
FEF	Iron Formation	41	39
FES	Iron Stone	27	21
AP	Ancient (Tertiary and older) Placer Au-U(Th)	34	33
MP	Modern (Quaternary) Placer Au-Th-Diamond	32	31
CMSD	Cyprus-type Massive Sulfide	68	65
SMSD	Superior-Type Massive Sulfide	104	73
EBAG	Epithermal Bulk Silver	51	46
PCD	Porphyry Copper	200	135
SMD1	Granite Porphyry Molybdenite (Stockwork Molybdenum, Type 1)	90	85
SMD2	Quartz Monzonite Porphyry Molybdenite (Stockwork Molybdenum, Type 2)	39	27
SKW	Tungsten Skarn	69	65
Totals:	23 Regional-Scale Models	1683	1370

However, during the three years when the nine prospect scale models were developed, we learned to recognize some common patterns in structuring the inference networks, and became more effective in the interviewing process. Equally important, we developed knowledge acquisition aids, such as the KAS system described in section 6, that greatly simplified knowledge-base development. As a result, all 23 of the regional scale models were encoded in less than a year. While much remains to be done to obtain complete coverage, we believe that the feasibility of the general approach has been well demonstrated.

5.2. Inference Network Compilation for Drilling-Site Selection

5.2.1. Background on Exploration Drilling. Mineral exploration projects typically go through a series of stages that can stretch over a period of many years. The early exploration phase may convince the geologists that they have a good prospect, but its real value cannot be determined until ore-grade mineralization has been found. For deep-seated ore bodies, this usually requires expensive diamond drilling, and considerable thought and effort are devoted to deciding where to drill.

To make this decision, the geologists may spend several months mapping the important properties — geological structures (such as faults and intrusions), rock types, alteration minerals, geophysical data, and geochemical data. Once the most important properties have been mapped, the problem is to combine all of this information to determine the drilling location that is most likely to intercept ore. If ore is found, subsequent drilling may be done to delineate its full extent, so that total grade and tonnage can be estimated.

Figures 4 and 5 show examples of some of the kinds of maps that are obtained. These maps give the surface expression of a buried ore body, and their proper interpretation again requires geological knowledge. In particular, it requires knowledge of the type of ore deposit that is expected to be present, which is the conclusion of the prospect evaluation study.

5.2.2. Inference Networks for Drilling-Site Selection. Superficially, this problem seems to be quite different from the prospect evaluation problem. However, we have developed a straightforward way to apply the very same inference network techniques to processing map data. The procedure is based on dividing the area into a large number of small cells. For each cell, the problem is to decide the degree to which that cell is a favorable location. Knowing the type of the ore deposit, one can construct an inference network that uses the mapped information to make that evaluation. The evidence is by no means restricted to that particular cell, but may be drawn from elsewhere on the map. By repeating the procedure for every cell, one can obtain a new favorability map that directly indicates appropriate drilling sites.

When PROSPECTOR is used in this application, the program begins by using the consequent mode to obtain information, not all of which is map data. When a map is requested, the program first asks if the desired map is available. If it is, the user enters it by using a digitizing tablet with a contour editing and display program; if it is not, the program will just have less information to work with.

After all of the available information has been entered, the program has the answers for every cell on the map. The only problem is that there can be a very large number of cells — 16,384 for the 128-by-128 resolution that we typically use. Even though only a few seconds are required for the interpreter to propagate the information through a typical inference network, this can lead to many hours of computation. Fortunately, the great speedup provided by the inference network compiler (described in section 4) reduces this to only a minute or two of CPU time.

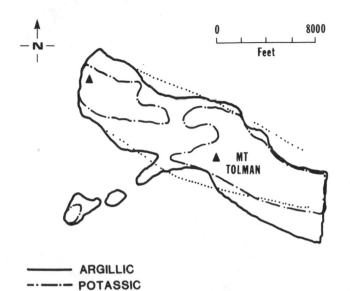

Figure 4. Geological maps of rock alteration for the Mt. Tolman deposit.

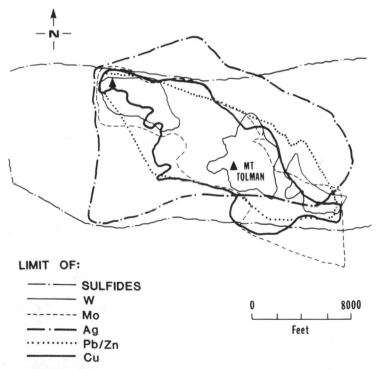

Figure 5. Geological maps of mineralization for the Mt. Tolman deposit.

As Table 2 indicates, three different drilling-site-selection inference networks have been developed. Like the other nets, they have been evaluated by testing them using data from well-known deposits. In addition, they were used in the only experiment in which PROSPECTOR was used to make a prediction on a prospect undergoing exploration. Since the results illustrate the method well, we shall present them briefly.[11]

The principal map data for this experiment are shown in Figures 4 and 5. These maps show the estimated boundaries of different types of rocks and metallic minerals for an area around Mt. Tolman in eastern Washington. This area had long been explored for base and precious metals, and a significant molybdenum deposit had been located and drilled by the company that provided us with these data in 1979; Figure 6 shows the company's estimate of the molybdenum ore body at that time.

Using only the 1979 information, in February, 1980, PROSPECTOR produced the favorability map shown in Figure 7.[12] Here the −5 to 5 favorability scale has been brightness coded, so that the most favorable spots are the brightest. In addition to showing agreement with the known mineralization, this map indicated the probable presence of ore-grade mineralization in the undrilled area to the northwest of the main ore body, and a barren zone between them. Subsequent drilling in the summers of 1980 and 1981

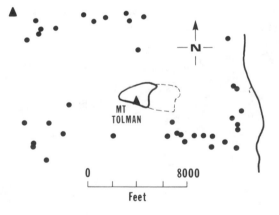

**◁⁖ KNOWN OUTCROP AND PROJECTION OF
PORPHYRY   MINERALIZATION, 1978
• BASE METAL, PRECIOUS METAL, AND
TUNGSTEN PROSPECTS, SMALL MINES**

Figure 6. *The size of the Mt. Tolman deposit estimated before PROSPECTOR's prediction.*

---

[11] A detailed report on this experiment is given in Campbell et al., (1982).

[12] This map was included in a project report (Duda, 1980) and published in a journal article (Duda & Gaschnig, 1981) before we were informed about the results of further exploration drilling in the area.

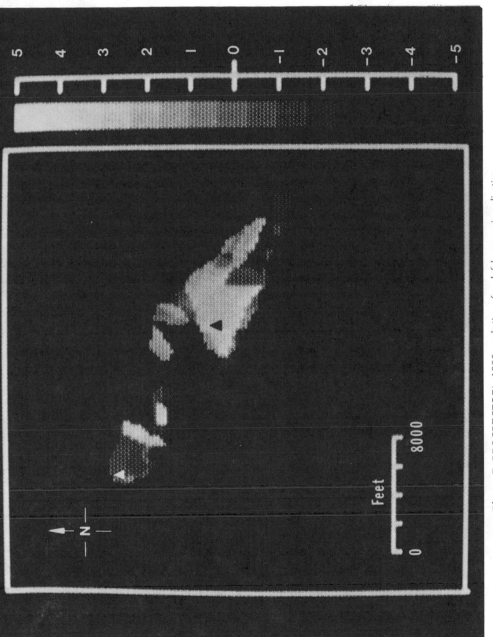

Figure 7. PROSPECTOR's 1980 *prediction of molybdenum mineralization.*

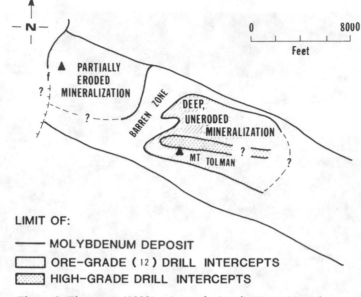

Figure 8. *The current (1982) estimate of mineralization at Mt. Tolman.*

substantially confirmed these predictions. Figure 8 shows the company's current estimate of the extent of ore-grade mineralization. Although this is only one test, it provides great encouragement for future use of these techniques.

5.3. Parameter Estimation — HYDRO

The representational and computational formalisms employed in PROS-PECTOR were originally developed for multiclass decision making. Here the basic problem is to make a selection from a usually small number of discrete alternatives. There are many other decision-making problems, however, in which accurate estimation of some quantity is required, such as in estimating the grade and tonnage of an ore deposit or the potential market size for a business organization.

We have addressed this problem in the area of estimating parameters for simulation models. Such models are widely used in such fields as me-teorology, hydrology, aeronautical engineering, and economics. One lim-itation of such simulation models is that the user must understand both the scientific basis of the domain and how that scientific knowledge is represented computationally in the model. Much judgment is often required in selecting parameters that determine the behavior of the system and in interpreting the results of a simulation run.

HYDRO is a PROSPECTOR-based parameter estimation system (Gaschnig, Reboh, & Reiter, 1981; Reboh, Reiter, & Gaschnig, 1982). Developed at SRI International, it is intended for use with the HSPF numerical simulation program developed by Hydrocomp, Inc., for the study of water resource problems. HYDRO advises users on the selection of values for "set up" parameters. These parameters indirectly describe watershed characteristics such as geography, vegetation, soil, slope, and geology.

Figure 9 illustrates some of the extensions to PROSPECTOR's inference network language that permit the representation of numerical computations. The simplest of these methods are look-up tables and algebraic formulas. If $E_1, E_2, \cdots E_n$ are nodes in the inference network that correspond to factors bearing on the computation of some quantity H, then the value of H can

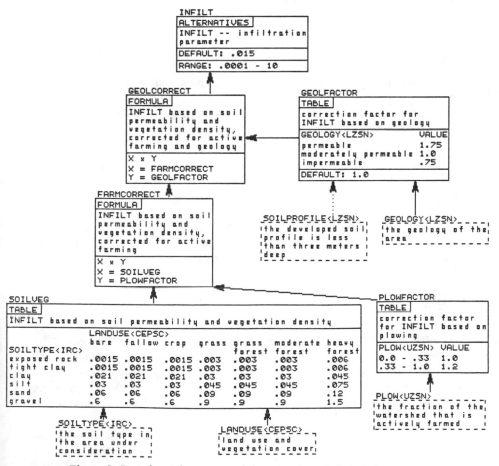

Figure 9. Part of an inference network for estimating a hydrological parameter.

be stored in an n–dimensional table. As in conventional table-lookup, the value associated with each of the n nodes serves as a key for searching the entries along the corresponding dimension in the table. However, if there is uncertainty in the values of the keys, one obtains a distribution in the possible values for H. Alternatively, one can specify an algebraic formula for computing the value of H from the values of the factors E_i taken as arguments (see Figure 9).[13]

During a consultation the user can express his or her uncertainty about an input quantity E_i by specifying a certainty factor for various intervals. Probability distributions are computed at each node H in the inference network by combining the probability distributions associated with its antecedents E_i. The exact combination methodology is determined by the type of numerical computation construct associated with H, and is described in Gaschnig et al., (1981) and Reboh et al., (1982).

6. KAS

It is widely recognized that the construction of a knowledge base is one of the most difficult and time-consuming parts of building an expert system. Just as the possession of a good programming environment simplifies the development of programs, the possession of good knowledge acquisition tools simplifies the development of knowledge bases.

KAS is the knowledge acquisition system that was developed to facilitate the construction and maintenance of PROSPECTOR's and HYDRO's knowledge bases. These two systems employ various kinds of networks to represent knowledge—*inference networks* for expressing judgmental knowledge, *semantic networks* for expressing the meaning of the propositions that correspond to nodes in the inference networks, and *taxonomic networks* for representing static knowledge about relationships among domain objects.

The core of KAS is an "intelligent" network editor that can assist the user in building, testing, searching, and maintaining these networks. Its basic operations allow it to create, modify or delete various kinds of nodes and arcs. It knows, however, about the representation constructs and inference mechanisms employed by PROSPECTOR and can therefore protect the user against certain kinds of syntactic errors. It also includes a bookkeeping system that keeps track of incomplete data structures. Whenever he or she desires, the user can turn control over to KAS, which will sys-

---

[13] Some other extensions to PROSPECTOR include a method for specifying alternative ways to compute or acquire the value of H, a method for allowing the user to override a value computed by the system, and a method for specifying conditions that must be satisfied by a value (Reboh et al., 1982).

tematically question him or her to fill in the missing parts of the structures. A *semantic network matcher* gives the user a limited ability to access the knowledge base by content rather than by form. The matcher also supports features such as protecting against numerical inconsistencies in the inference networks, generating meaningful explanations, and enhancing the communication between the user and the system. Finally, because KAS contains PROSPECTOR's inference mechanism as one of its components, it permits controlled execution of individual sections of an inference network, enabling the knowledge engineer to monitor his or her progress in refining the knowledge base. Detailed description of KAS can be found in Reboh (1981).

7. Discussion

We have discussed the principles and some of the applications of inference networks — PROSPECTOR's central method of representing knowledge for decision making. While the details can become complex, the basic concepts are quite straightforward. Inference networks provide a simple and natural structure both for expressing the factors that go into making a decision and for specifying how degrees of belief in these factors are to be combined to reach a conclusion. Throughout the paper, we have taken the position that the expert is the one who should specify how beliefs are to be combined, and that inference networks effectively provide a formal language for that task. Thus, we would like to conclude with some general observations on the features of formal languages in general.

The development of high-level languages marked a major advance in computer science. Although a programmer still describes knowledge about how to solve a problem by specifying algorithms or procedures, a high-level language allows the algorithmic considerations to be separated from the details of the instruction set for a particular machine. The use of formal knowledge representation languages can be viewed as another step in this direction (Barr & Feigenbaum, 1981). Here the programmer does not specify exactly how the knowledge is to be used, but merely describes the necessary knowledge. Thus, the builder of the knowledge base for PROSPECTOR is still programming, but is using inference networks as a language for expressing the information to be used by the problem solving procedures.

Viewed broadly, mathematical theories such as decision theory and formal logic can also be viewed as knowledge representation languages. Statistical decision theory provides a very general framework for the treatment of decision problems involving uncertainty. However, as we mentioned earlier, there is good reason to doubt that it corresponds well to methods normally employed by people. To use this language, one must express knowledge of the problem in terms of joint probability functions; aside from a few general concepts such as statistical independence and Bayes's rule, the theory does not provide any helpful methods for structuring one's knowledge.

Formal logic provides quite a different framework, one capable of arbitrary structural complexity. Compared to formal logic or other general-purpose languages, inference networks are quite limited in their expressive power; however, they are more immediately applicable to decision problems. Compared to probability theory, inference networks are not as well founded mathematically; however, they provide a greater variety of standard methods for expressing known relations. Although they are not universally applicable, we have found them to be very effective for solving interesting and important judgmental decision problems.

8. References

Barr, A., & Feigenbaum, E. A. (Eds.). *The handbook of artificial intelligence* (Vol. 1). Los Altos, CA: Wm. Kaufmann, 1981.

Buchanan, B. G. Research on expert systems. In J. E. Hayes, D. Michie, & Y. H. Pao (Eds.), *Machine intelligence 10.* New York: Wiley, 1982.

Campbell, A. N., Hollister, V. F., Duda, R. O., & Hart, P. E. Recognition of a hidden mineral deposit by an artificial intelligence program. *Science,* 1982, *217,* 927–929.

Davis, R., & King, J. An overview of production systems. In E. W. Elcock and D. Michie (Eds.), *Machine intelligence 8.* New York: Wiley, 1976.

Duda, R. O. *The Prospector system for mineral exploration.* (Final Report, SRI Project 8172). Menlo Park, CA: SRI International, Artificial Intelligence Center, April 1980.

Duda, R. O., & Gaschnig, J. G. Knowledge-based expert systems come of age. *Byte,* 1981, *6,* 238–281.

Duda, R. O., Hart, P. E., Konolige, K., & Reboh, R. *A computer-based consultant for mineral exploration* (Final Report, SRI Project 6415). Menlo Park, CA: SRI International, Artificial Intelligence Center, September 1979.

Duda, R. O., Hart, P. E., & Nilsson, N. J. Subjective Bayesian methods for rule-based inference systems. *Proceedings of the National Computer Conference,* 1976, *45,* 1075–1082.

Duda, R. O., Hart, P. E., Nilsson, N. J., Reboh, R., Slocum, J., & Sutherland, G. L. *Development of a computer-based consultant for mineral exploration* (Annual Report, SRI Projects 5821 and 6415). Menlo Park, CA: SRI International, Artificial Intelligence Center, October 1977.

Duda, R. O., Hart, P. E., Nilsson, N. J. & Sutherland, G. N. Semantic network representations in rule-based inference systems. In D. A. Waterman and F. Hayes-Roth (Eds.), *Pattern directed inference systems.* New York: Academic Press, 1978.

Duda, R. O., & Shortliffe, E. H. Expert systems research. *Science,* 1983, *220,* 261–268.

Gasching, J. Application of the PROSPECTOR system to geological exploration problems. In J. E. Hayes, D. Michie, & Y–H Pao (Eds.), *Machine intelligence 10.* New York: Halsted Press, 1982.

Gaschnig, J. G., Reboh, R., & Reiter, J. *Development of a knowledge-based expert system for water resource problems* (Final Report, SRI Project 1619). Menlo Park, CA: SRI International, Artificial Intelligence Center, August 1981.

Hart, P. E., Duda, R. O., & Einaudi, M. T. Computer-based consultation system for mineral exploration. In A. Weiss (Ed.), *Computer methods for the 80's in the mineral industry.* New York: Society of Mining Engineers, 1979.

Konolige, K. An inference network compiler for the Prospector rule-based consultation system. *Proceedings of the Sixth International Joint Conference on Artificial Intelligence,* 1979, 487–489.

Nau, D. S. Expert computer systems. *Computer,* 1983, *16,* 63–85.

Pople, H. Heuristic methods for imposing structure on ill-structured problems: the structuring of medical diagnostics. In P. Szolovits (Ed.), *Artificial intelligence in medicine.* Boulder, CO: Westview Press, 1982.

Raiffa, H. *Decision analysis: Introductory lectures on choices under uncertainty.* Reading, PA: Addison-Wesley, 1968.

Reboh, R. Using a matcher to make an expert consultation system behave intelligently. *Proceedings of the First National Conference on Artificial Intelligence,* 1980, 231–234.

Reboh, R. *Knowledge engineering techniques and tools in the Prospector environment* (Technical Note 243). Menlo Park, CA: SRI International, Artificial Intelligence Center, June 1981.

Reboh, R., Reiter, J., & Gaschnig, J. G. *Development of a knowledge-based interface to a hydrological simulation program* (Final Report, SRI Project 3477). Menlo Park, CA: SRI International, May 1982.

Shortliffe, E. H. *Computer based medical consultations: MYCIN.* New York: American Elsevier, 1976.

Shortliffe, E. H., & Buchanan, B. G. A model of inexact reasoning in medicine. *Mathematical Biosciences,* 1975, *23,* 351–379.

Shortliffe, E. H., Buchanan, B. G., & Feigenbaum, E. A. Knowledge engineering for medical decision making: A review of computer-based clinical decision aids. *Proceedings of the IEEE,* 1979, *67,* 1207–1224.

Waterman, D., & Hayes-Roth, F. (Eds.) *Pattern directed inference systems.* New York: Academic Press, 1978.

Zadeh, L. A. Fuzzy sets. *Information and Control,* 1965, *8,* 338–353.

Using Expert Systems to Manage Change and Complexity in Manufacturing

Dennis E. O'Connor
SENIOR GROUP MANAGER
INTELLIGENT SYSTEMS TECHNOLOGIES GROUP
SYSTEMS MANUFACTURING
AI TECHNOLOGY CENTER

DIGITAL EQUIPMENT CORPORATION
77 REED RD.
HUDSON, MA 01749

Increasing productivity and being competitive in the 1980s in manufacturing requires Digital to pursue new and innovative methods and tools to help manage and solve problems in a very complex environment.

1. Introduction

Digital has established industrial leadership worldwide in the use and development of state-of-the-art Artificial Intelligence methodologies (expert systems) applied to real complex problems in the manufacturing environment. In collaboration with Carnegie-Mellon University's Computer Science Department, Digital has established an effective technology transfer process to aid in the joint development and design of new innovative Intelligent System tools. The effective use (by actual managers) of these expert systems will have a significant impact on productivity goal achievement, and ability to manage changes in a complex manufacturing business environment.

2. The Pioneering Effort at Digital in AI

The acknowledgement that Artificial Intelligence methodologies had the potential of aiding managers in solving complex manufacturing problems became apparent during Digital's Systems Manufacturing pioneering effort in AI during the fall of 1979.

This effort involved the development of our expert system known as XCON, an Expert Configurator of VAX 11780 Computer Systems. Until the availability of XCON, the computer system configuration task in manufacturing was accomplished manually, an extremely difficult task, requiring many knowledgeable resources. Traditional methodologies did not yield the accurate, cost-effective, fast response that the new XCON Expert System Configurator provided.

The majority of the computer system orders that customers send to Digital exhibit a large degree of à la carteness (many line items per order). The development of XCON represented a very significant tool for managing the complexity of Digital's business. Prior to the availability of XCON, technical editors in manufacturing reviewed all customer orders for technical correctness and order completeness. A dramatic increase in productivity of the technical editors has occurred due to the effective use of the XCON expert system in the manufacturing environment.

The XCON Expert System is implemented in OPS–5, which is a general purpose rule-based language. Like other rule-based languages, OPS–5 provides a rule memory, a global working memory, and an interpreter that tests the rules to determine which ones are satisfied by the descriptions in working memory. An OPS–5 rule is an IF–THEN statement consisting of a set of patterns which can be matched by the descriptions in working memory and a set of actions which modify working memory when the rule is applied. On each cycle, the interpreter selects a satisfied rule and applies it. Since applying a rule results in changes to working memory, different subsets of rules are satisfied on successive cycles. OPS–5 does not impose any organization on rule memory; all rules are evaluated on every cycle. When more than one rule is satisfied on a cycle, the interpreter uses a few general conflict resolution strategies to decide which rule to apply.

2.1. Increased Productivity and Manufacturing Cost Savings

When the Expert System XCON was first installed in manufacturing, the system had limited capability and knowledge. XCON was about equal in performance to the manual technical editor process. Therefore, much interaction was required with the technical editors to increase the expertise of this system. Traditionally trained technical editors in manufacturing required 20–30 minutes per system order to configure each system order. This robust expert system contains over 2000 rules and configures extremely complex system orders in under *a minute*. XCON provides additional functions and capabilities not formerly performed by the traditional technical editors such as (a) defining the exact cable length for all cables required between each system component, and (b) it provides the vector addresses calculation for the computer bus options.

Today, all VAX family system orders are configured by XCON in the United States and European plant operations. As the workload has increased in the system manfuacturing plants, additional technical editors have not been added to the workforce, due to the increase in productivity and capacity provided by the XCON system. In the past, a large portion of the System orders, being scheduled to the factory floor, had numerous configuration errors and lacked completeness; now, with the use of XCON, thousands of VAX orders scheduled to be built have very accurate configurations 98% of the time. Benefits derived in the manufacturing operations from having accurate System configurations are: (a) an increased thru-put order rate, (b) few shipments are delayed due to System configuration errors, (c) better utilization of materials on hand.

When numerous line item changes occur on customer orders already scheduled in the manufacturing process, the effective use of the XCON system provides a tool to save time, increase output per person, and lower the manufacturing cost in this complex task of computer assembly, as customer change orders arise. XCON generates a configuration sheet which is used by the technician on the factory floor to help configure the VAX 11780, 11750, 11730 systems. Technician productivity has increased substantially, due to the availability of this accurate system configuration information. In the past, numerous hours in computer assembly were spent determining the proper cable lengths required on each System order.

Better utilization of Digital's people assets due to the redeployment of highly skilled senior technicians has now occurred, allowing them to address more technically difficult tasks. Cost savings to Digital of hundreds of thousands of dollars are occurring now, and will continue through the 1980s. The plant management has been extremely satisfied with the emerging Artificial Intelligence methodologies, and they too have been participating in this pioneering effort with much enthusiasm.

3. Why Digital Chose Expert Systems

Digital has always offered its customers a very wide range of computer system products. The range of product offerings span the micro PDP–11 to the very large systems in the VAX family and PDP 10/20 series. Since Digital has essentially offered an à la carte menu of system product options, it is not uncommon to have thousands of customer orders sent in, each one being somewhat different. This à la carte product offering creates, therefore, a very complex business environment. This environment creates a complex series of events which involves processing the orders, scheduling manufacturing to build the orders, and lastly to distribute computer system orders to the customer in a timely manner.

Digital Manufacturing Business, then, exhibits some of the characteristics where expert systems could be brought to bear to aid managers in managing complexity and change in their operations. The XCON (Expert VAX Systems Configurator) is just one example of complex task domain, where expert systems have been used at Digital. We essentially looked to the AI methodologies of expert systems as a new and innovative way of helping manage *change*, complexity, and exception management in the Digital business operations.

As Digital business grows, and new market opportunities emerge and major changes occur, an increase in the business complexity is expected. This new way of conducting business, with aggressive productivity goals and asset goals, requires the use of innovative, highly responsive information management systems.

Expert systems will play a major strategic role in helping DEC managers effectively manage the change of business and the complexity of their operations. Further, expert systems will coreside with the excellent traditional management information systems already in operation, in the major functions of sales, engineering, manufacturing and field services.

Although we are just learning how to apply these new Artificial Intelligence methodologies to complex business problems, our current experience strongly suggests that these intelligent systems are possibly the best methodology for this class of dynamically changing problems. Capturing the knowledge and heuristics from the manufacturing operations experts and distribution logistic experts will provide a powerful knowledge base that will aid the management of these complex business domains. New manufacturing operations managers will have access to this expert knowledge base and thus their learning curves should ramp dramatically.

Expert systems provide for managing change, in a less disruptive way, since the rule-base system can be modified quite easily, even on a daily basis. As events and business strategies require updating, these new rules can be added without causing major upheavals within the business operations.

Traditional management information systems that require major changes usually cause a disruptive effect in the business operations, especially when major functionality must be changed.

Early involvement of the end user in the start up process of expert systems development has helped the migration of this technology tremendously. Since expert systems are intensely user-driven intelligent system tools, the end user becomes involved from the beginning of the framing of the problem to prototype shell build, continuing through the knowledge acquisition process, to the final product.

We have observed a growing synergistic relationship between the Intelligent Tool and the people who design and then use them. As more and more of a person's heuristic knowledge is encoded into these programs, the more these people are attached to the program usage. Some of the end users of these expert systems view the capability the expert system provides as having an amplifier effect on their abilities to perform more effectively in their jobs.

3. The Problem Domain

The key to effective use and development of intelligent system tools depends on being able to identify the "core problem space," and the bounded framework of that problem domain. This framework must have an identifiable linkage to the business goals. This business goal linkage is necessary to ensure a focused effort occurs, resulting in a substantial return on our investment. In general, most of the current Expert Systems under development in Digital manufacturing are linked to a major business transformation, or a major people productivity improvement.

Some characteristics the manufacturing problem domains exhibit are:

- *Ill-Structuredness.* When complete data is not available. The configuration task (VAX 11780) exhibited this ill-structuredness, because all the information and data were not available at the start with respect to how all the components should be interconnected.
- *Key Parameters.* Change on a daily or weekly basis. In the order scheduling task, changes occurred due to daily modification of the different scheduling strategies in each of 18 different product line businesses.
- *Interdependencies.* Complex interconnect of organizations. The interdependency characteristic is exhibited in the complex logistics management distribution problem. Shipping and distributing product from over 30 plants must depend on the logistical connection among business centers, plus Sales and Field Service organizations.
- *The "Jello" Effect.* The manufacturing environment constantly changes due to: changes in business input, increase in orders, order cancellation, changing the mix of inventory required to meet new customer demand. Traditional algorithmic solutions can't handle "change demand" due to business increase or decrease. There are other characteristics that the problem domain should exhibit, but the four described here have some of the predominant properties that any "core problem space" should possess, when deciding to use the Expert System methodologies.

4. Intelligent Systems Applications in Manufacturing

4.1. The Intelligent Scheduling Assistant, ISA

ISA is a rule-based system that provides the capability of scheduling customer system orders against the current and planned material allocations. ISA is currently in use at Digital System manufacturing plants as an extension to Digital's Corporate Common Scheduling System. Although still in the development phase, ISA has been actively used to schedule customer orders for the past year and a half. Expert schedulers continue to input new rules and knowledge to this system. ISA's performance as a scheduling assistant has attained a scheduling rate of "orders per second." Traditional manufacturing order schedulers require at least 10–15 minutes per system to perform this task.

ISA's current version has the following capabilities:

- Each night it reads two files produced by the Common Scheduler containing the set of transactions processed by the Common Scheduler that day. One file contains cancellations, and the other contains the remaining transactions.
- After dealing with all cancellations, it tentatively schedules all new orders and all change orders.
- Those orders that it can tentatively schedule in the month desired by the customer (without applying any of its loading strategies) are designated as schedulable, and the month that each of these orders is to be scheduled is written to a file. This file is then passed to the Common Scheduler and the orders are scheduled. ISA also produces a report containing a one line notification of the month in which each order passed to the Common Scheduler has been scheduled.
- Those orders that ISA must use its loading strategies on fall into two classes: (a) orders that are problematic because of a material shortage, and (b) orders that are problematic because of a credit line insufficiency. ISA tentatively schedules each of these orders in a month that satisfies the various loading constraints that it knows about. (New ones continue to be added). Then in an effort to schedule closer to the desired date, ISA tries again with the loading constraints relaxed. If a scheduling is found which is closer to the demand date, ISA notes it as an alternative. It then produces two reports, one containing a one-line proposal for when to schedule each order which has material problems and the other containing a one-line proposal for when to schedule each order that is problematic because of a credit line insufficiency.

- After the reports are printed, ISA can be used to assist in the scheduling of the problematic orders. When ISA is run, it will prompt for the DEC number of a problematic order: when a DEC number is entered, it will display a screenful of information indicating when it proposes the order to be scheduled, and what the problems of scheduling the order are. If ISA has alternative proposals for when the order might be scheduled, it will display those as well. ISA employs a number of strategies to determine how to select a scheduling date. Other strategies are concerned with partial shipments, roll forward by quarter, suggested substitutions, and speculative orders.

Today ISA has gained substantial expert knowledge, and effective scheduling strategies. It is to be considered as the next generation scheduling system to replace Digital's Corporate Common Scheduling System, which handles all of Digital's System orders.

4.2. *The Intelligent Management Assistant for Computer Systems Manufacturing*, IMACS

IMACS is currently under development and in use in a Systems manufacturing plant. It provides for the 4-wall manufacturing environment the capability to assist managers in paperwork management, capacity planning, floor-loading thruput management, diagnostic selectors, and inventory management.

IMACS' domain of expertise is computer system manufacturing. Its knowledge of how to assist the management of the computer systems manufacturing process is encoded as rules, each of which defines a situation in which it is relevant and an action which is appropriate in that situation. IMACS is implemented as a set of cooperating knowledge-based programs. Each of these subsystems is an "expert" in one aspect of the System manufacturing (FA&T) plant management task.

IMACS' primary focus is on helping to manage exceptions, and providing assistance to problems that arise in the manufacturing process. The IMACS system provides the following kinds of help:

- It provides assistance to the people responsible for scheduling by constructing a build plan for each system, determining when the system should begin to be built and where it should be placed on the floor, and monitoring the build plan if it becomes unimplemental because of resource or time constraints.
- It provides assistance to the people responsible for acquiring materials by drawing their attention to possible impending shortages or surpluses.

- It provides assistance to people responsible for dealing with order administration issues by indicating the dates by which those issues must be resolved.
- It provides general assistance by generating reports summarizing plant performance along various dimensions.

The input that the IMACS operates on is the flow of customer orders from the order-processing group. The customer order has already been assigned a delivery date. It waits to construct a detailed build plan, since initially only an outline build plan needs to be created. IMACS creates an abstract build plan which enables it to roughly estimate the resource requirements for the order, so that if the plant's projected resource requirements are significantly off, it can alert someone that problems may arise. Several weeks before the system is to be built, IMACS constructs a detailed plan and then monitors implementation of that plan.

The IMACS system is just starting in its development phase. Many man hours need to be worked to build this system into an effective management tool for a 4-wall plant environment.

5. Some Lessons Learned

Digital at this point has 4½ years experience using Expert Systems routinely on a daily basis in United States and European manufacturing plant operations. During this time we have crept up the learning curve, making mistakes as we proceeded, but always learning. We have internally developed resources in AI to service the current applications under development. Further, we have established an internal Digital AI Training Center to help meet our growing needs in the Expert Systems space.

As we make progress using Expert Systems at Digital, a few lessons are worth mentioning. I believe these eight points make the difference in establishing an effective technology transfer process for AI methodologies:

- The assessibility to senior management who nurture, sponsor and guide innovation is a must.
- Identify and bound your "core" problem space.
- Ensure a linkage to business goals.
- Demonstrate your Expert System (shell) early.
- Manage cultural resistance and management expectations persistently.
- Where possible, provide seed funding for new systems.
- Manage the technology transfer: have a plan.
- Establish educational/training capability.

6. Summary

The *pioneering effort* of developing applications in Artificial Intelligence *continues* within Digital's sales groups; manufacturing groups; engineering groups; field service groups, and Artificial Intelligence Technology Center.

Although the emphasis in this paper has been on Manufacturing, these other groups are actively involved in this pioneering effort. The potentiality of the field of Artificial Intelligence is indeed vast. Digital's main use to date of these state-of-the-art methodologies has been to aid in making the IL (indirect labor) more productive, and the complex problems more tractable, when encountered by the manufacturing management.

The Industrialization of Knowledge Engineering

Frederick Hayes-Roth
CHIEF SCIENTIST
TEKNOWLEDGE INC
Palo Alto, California 94301

For the past 15 years, applied work in artificial intelligence has focused increasingly on the use of knowledge to build "expert systems." These systems achieve levels of performance in complex tasks that equal or even exceed that of human experts. Because they incorporate much human knowledge, these systems are called knowledge-based expert systems or, simply, knowledge systems. The subfield of applied artificial intelligence that addresses the design and development of knowledge systems is called knowledge engineering.

The industrialization of knowledge engineering began in 1981 with the formation of two commercial spin-offs from the Stanford University Heuristic Programming Project. One of these focuses exclusively on applications of knowledge engineering to genetics research. The other, Teknowledge, focuses on industrial and commercial uses of knowledge engineering. It designs and implements high-valued systems for a number of clients both in the US and in Europe. Sales this year will be $3 million to $6 million. From my perspective as the chief technologist in Teknowledge, I would like to survey for you the technology and economics of knowledge engineering.

To begin, I would like to place knowledge engineering in context, at least as we see it relating to AI in general. Figure 1 surveys the major developments in the field of AI over its 25-year history. To understand the field's evolution, you should grasp the major themes shown at the top of six columns.

A BRIEF HISTORY OF AI

CONCEPTION	REDIRECTION	SUCCESS	RECOGNITION	INITIATIVE	INDUSTRIALIZATION
GPS	DENDRAL	MYCIN	TI	Japan, Inc.	Xerox
LT	Robots	META-DENDRAL	EMYCIN	Teknowledge	HP
IPL	Theorem Proving	HEARSAY-II	U.S. Air Force	PROSPECTOR	DEC
Symbolic Programming	Symbolic Math	MACSYMA	AT&T	U.S. Navy	GE
Brute Force	LISP	HARPY	VLSI	DSB	Martin-Marietta
Generality	Heuristics	REDUCE	Xerox	R1	TRW/ESL
	Satisficing	INTERLISP	Symbolics	LMI	E-Systems
	Specialization	IF-THEN Rules	PROLOG	ROSIE	Elf
		Knowledge Engineering	Nobel Prize	KS300	NCR
		Domain Knowledge	Schlumberger	Fairchild	Amoco
					XSEL
					S1 LISP
					Business Week
					Fortune
					Technology
					Wall St. Journal
					Time
					Newsweek
1956	**1968**	**1977**	**1980**	**1981**	**1982**

Figure 1.

1. *History of* AI

- Conception. In 1956, the founders of the field hoped that symbolic computing, coupled with powerful general purpose computers, would solve important and complex problems.
- Redirection. By 1968, most workers had abandoned their initial conception as naive and redirected their efforts towards knowlege-based reasoning and heuristic methods.
- Success. By 1977, independent observers judged the knowledge-based methods successful, as major scientific and technical hurdles were crossed.
- Recognition. By 1980, many industrial and government leaders finished their own evaluations of the technology and publically certified the technology as crucial and relevant.
- Initiative. In 1981, major new initiatives occurred ranging from the Japanese MITI's Fifth Generation Machine project to the formation of Teknowledge by 20 leading knowledge engineers from Stanford, Rand and MIT.
- Industrialization. Beginning in 1982 and proceeding now at a quickened pace, many companies adopted the technology to solve current problems and the press recognized the emergence of the field.

Figure 2 summarizes the important lessons we have learned from this evolutionary history. The major point concerns the central importance of knowledge in these AI applications. Human know-how makes natural systems work, and knowledge engineers make artificial systems perform by equipping them with similar know-how. Unfortunately, humans don't possess a neat codification of their know-how. Knowledge engineers must interrogate knowledgeable people to elicit this know-how, and then they must model what their informants reveal. This process leads to what is often the very first written codification of human knowledge. The initial for-

---

LESSONS LEARNED OVER 25 YEARS OF AI EXPERIENCE

- Knowledge powers the solution to important problems
- Knowledge resides in unrefined chunks
- A knowledge system can comprise 100, 100,000 or 100,000,000 chunks
- The key tasks: Mining, refining, molding and assembling knowledge
- These tasks give rise to knowledge engineering

Figure 2.

mulation is rarely correct, of course, and the knowledge engineering team iteratively reformulates the knowlege system as the informants come to understand what they know.

Figure 3 elucidates the key tasks in knowledge engineering and defines their products. The engineering products may sound quite abstract to you, although they are quite concrete and specific in the knowledge engineering discipline. By interviewing experts, the knowledge engineers acquire their knowledge and represent it in the form of specific concepts and rules. These form the lexicon and contents of the knowledge base. To organize this unstructured base of knowledge, the knowledge engineers must design an overall framework and systematic representation scheme. Given this framework, the knowledge engineer then assembles the individual components of knowledge into an organized base of rules for a specific inference engineer to interpret and apply. Finally, because no initial body of rules ever achieves expert performance, the errors and omissions revealed by the knowledge system stimulate the team to revise the knowlege base.

I would like to point out six of the most common jobs that companies currently purchase knowledge systems to perform. These are shown in Figure 4. I will clarify each of these functions by illustrating the types of

KEY TASKS IN KNOWLEDGE SYSTEM DEVELOPMENT

KNOWLEDGE PROCESSING TASKS	ENGINEERING ACTIVITIES	ENGINEERING PRODUCTS
Mining	Knowledge acquisition	Concepts and rules
Molding	Knowledge system design	Framework and knowledge representation
Assembling	Knowledge programming	Knowledge base and inference engine
Refining	Knowledge refinement	Revised concepts and rules

Figure 3.

problems that motivate knowledge systems of these types. In these following illustrations, I will draw heavily on my industrial experience. However, the cases I will present combine properties from numerous actual cases in analogies that protect proprietary information. Because each case is self-explanatory, the figures are presented without further comment.

WHAT KNOWLEDGE SYSTEMS CAN DO

1. Solve problems that thwart data processing techniques
2. Preserve perishable expertise
3. Distribute expertise
4. Fuse multiple sources of knowledge
5. Convert knowledge into a competitive edge
6. Alter business common sense and perception

Figure 4.

COMPOSITE CASE STUDY #1:

MEMO

XYZ COMPUTERS, INC.

From: Executive VP, Computer System Products
To: Vice President, Marketing
Subject: Training Deficiencies

Please advise me on ways to improve sales training. Specifically, I want all systems ordered in complete and correct detail to prevent field engineering delays at installation.

Figure 5.

MEMO

XYZ COMPUTERS, INC.

From: Vice President, Marketing
To: Executive VP, Computer System Products
Subject: Limits to Training

Impossible! We are already a full six months behind in training our sales force on these complicated and rapidly changing products.

I suggest you tell the field engineering people to compensate for our minor sales errors.

Figure 6.

MEMO

XYZ COMPUTERS, INC.

From: Executive VP, Computer System Products
To: Vice President, Marketing
Subject: Checking Sales Orders

Please estimate for me the cost of a program to validate computer system sales orders for configuration validity and completeness.

Figure 7.

MEMO

XYZ COMPUTERS, INC.

From: Vice President, Data Processing
To: Executive VP, Computer System Products
Subject: Computer Order Checker

We are unable to computerize computer order checking. We have tried several times with significant efforts. Engineering changes components, features, parts, and software capabilities too often.

I strongly suggest you train the sales reps to do this task better.

Figure 8.

KNOWLEDGE SYSTEM SOLUTION #1:

ORDER TAKER AND CHECKER

KNOWLEDGE BASE

Engineering Configuration Rules	Local Installation Rules	Manufacturing and Inventory Rules	Delivery Rules

—Aids marketing in formulating orders
—Validates or corrects orders
—Gives feedback to sales person
—Generates MRP (Manufacturing Resource Planning) inputs

Figure 9.

✔ The Lesson of Case #1:

Knowledge Engineering Can Solve Problems of Great Economic Value Which Resist Solution by Other Methods.

COMPOSITE CASE STUDY #2:

MEMO

JKL POWER SYSTEMS CO.

From: Director Turbine Maintenance Division
To: Executive VP, Engine Products
Subject: Backlogged A100-Engine Repairs

Our A100-engine repairs backlog arises from sales exceeding our support capabilities.

We have only two maintenance engineers who can handle this line competently.

Incidentally, both engineers will retire within the next 24 months.

Figure 10.

KNOWLEDGE SYSTEM SOLUTION #2:

EQUIPMENT FAILURE DIAGNOSIS AND REPAIR

KNOWLEDGE BASE

Diagnosis   Therapy   Documentation
Rules   Rules   Rules

—Centralizes company knowledge of maintenance
—Speeds central diagnosis and repair
—Generates breakdown reports automatically

Figure 11.

✔ **The Lesson of Case #2:**

Knowledge Engineering Can Capture and Preserve Perishable Expertise.

COMPOSITE CASE STUDY #3:

MEMO

F & M INDUSTRIES, OIL SERVICES DIVISION

From: Executive Vice President, Oil Services
To: Chairman of the Board, F&M Industries
Subject: Problematic division revenue projections

The draft revenue projections greatly overstate our likely performance.

The assumed 30% growth rate requires adding 750 drilling engineers per year, while we can hire and train at most 125.

Losses attributable to poor judgments by the inexpert engineers we now employ already surpass $100,000,000 annually.

Figure 12.

KNOWLEDGE SYSTEM SOLUTION #3:

MEASUREMENT-WHILE-DRILLING ADVISOR

KNOWLEDGE BASE

Geological Formation Rules	Drilling History Rules	Dragging Analysis Rules	Sticking Problem Rules	Kick and Kill Rules

Figure 13.

✔ **The Lesson of Case #3:**

Through Knowledge Systems, Companies Can Syndicate Their Knowledge to Sustain Growth.

COMPOSITE CASE STUDY #4:

MEMO

IJK FOODS CO.

From: Vice President, Engineering
To: CEO
Subject: Building cost over-runs

We have completed the analysis of construction cost over-runs.

Engineering errors account for 80% of the over-runs. Of these, 10% were errors of "commission" and 90% were errors of "omission."

We need a way to collect and organize our vast experience so we can reuse it when designing new plants.

Figure 14.

KNOWLEDGE SYSTEM SOLUTION #4

DESIGN COST ESTIMATOR

KNOWLEDGE BASE

Normal Cost Factor Rules	Regional Variation Rules	Delay Rules	Design Rules

Real Estate
Manufacturing
A&E
Construction
Staffing
Furnishing
Tooling

Figure 15.

✔ The Lesson of Case #4:

Knowledge Engineering Enables You to Combine and Apply Routinely the Expertise of Many Different Specialists.

COMPOSITE CASE STUDY #5:

MEMO

BHP INSTRUMENTS, INC.

From: Division Director, Medical Instruments
To: Marketing Director
Subject: EKG Analysis Products

Your claim that a 10% improvement in performance could increase market-share by 50–100% prompted a major program review.

All products including those of our competitors perform about 75% correct analyses. Everything technically possible is already being done.

Unless you can tell us how to package a medical expert with every instrument, we're not going to do much better.

Figure 16.

KNOWLEDGE SYSTEM SOLUTION #5:

EKG INTERPRETER

KNOWLEDGE BASE

Patient History Rules	Statistical Analysis Routines	Disorder Rules	Anomaly Rules

—Combines medical expertise with signal processing skills
—Produces an MD's report, ready for signing
—Achieves 85% performance

Figure 17.

✔ **The Lesson of Case #5:**

Knowledge Engineering Converts Expertise Into a Competitive Advantage.

COMPOSITE CASE STUDY #6:

MEMO

ABC GEOLOGY, INC.

From: Comptroller
To: Chairman
Subject: Discounts and write-offs

Here's the answer to your question. Our customers reject as unsatisfactory 5–10% of all analyses we now produce.

Headquarters apparently employs a brilliant Geophysical Engineer to mediate complaints and authorize write-offs.

I estimate these discounts total $40 million per year. I suggest we run a Quality Achievement campaign to heighten awareness.

Figure 18.

KNOWLEDGE SYSTEM SOLUTION #6:

QUALITY ASSURANCE MONITOR

KNOWLEDGE BASE

Collection	Instrument	Anomalous
Analysis	Characteristics	Performance
Rules	Rules	Rules

—Enforces company quality procedures
—Validates data calibrations and ranges
—Rationalizes anomalous results

Figure 19.

✔ **The Lessons of Case #6:**

The Capacity to Engineer Knowledge Should Alter the Manager's Perception of Business Reality.

Often What Management Sees as a Problematic Fact-of-Life Turns Out to Be Just a Routine Task for Knowledge Engineering.

Figure 20 shows a list of areas in which knowledge engineers today are working to build knowledge systems. There are numerous areas in addition which are not shown. And in several of these areas, more than one company or laboratory now operates.

To illustrate one of these systems concretely, I have prepared some figures on one system Teknowledge has developed for Elf-Aquitaine, the French national oil company. They contracted our services to develop a knowledge system that could advise their drilling rig supervisors how best to overcome and subsequently avoid sticking problems. An average daily cost of each rig is roughly $100,000, so sticking is a very costly problem. Elf estimates it should save tens of millions of dollars per year through reduced down-times as a direct result of this knowledge system.

Figure 21 shows the view of the CRT screen which the drilling adviser presents to its user. On the left-hand side of the screen, we see elements of the adviser that pertain only to drilling and sticking. On the other side, we see components that reveal the inner workings of the system and its

CURRENT KNOWLEDGE ENGINEERING APPLICATIONS AREAS:

- Medical diagnosis and prescription
- Equipment failure diagnosis
- Computer configuration
- Chemical data interpretation and structure elucidation
- Experiment planning
- Speech and image understanding

- Signal interpretation
- Mineral exploration
- Military threat assessment and targeting
- Crisis management
- Advising about computer system use
- Training and teaching
- VLSI design

Figure 20.

ANATOMY OF A KNOWLEDGE SYSTEM: THE DRILLING ADVISOR

DOMAIN-SPECIFIC			GENERIC
User-Advisor Dialogue			Context Tree
			Goal Tree
Sticking Problem Hypotheses	Well Formation	Bottom Hole Assembly	Current Rule
			Inferred Conclusions

Figure 21.

reasoning processes. These capabilities are generic in that the same forms and methods appear in any knowledge system constructed with the same framework and underlying knowledge engineering tools.

The domain-specific elements of the drilling advisor consist of a dialog window where the advisor and the supervisor discuss current situations. Below the dialog window, three additional windows appear. The first of these graphically depicts the six hypotheses regarding the likely causes of sticking. Next, the drilling advisor depicts its understanding of the subterranean strata in which the well lies. Finally, the last window pictures the roughly two-dozen collars, devices, and bits that compose the current bottom hole assembly.

The current advisor has about 250 rules in it, which represents perhaps 90% of the total knowledge required to handle sticking problems and about 10% of the total knowledge required to handle all drilling problems the advisor will eventually cover. To illustrate, rule 121 is from the Drilling Advisor. This rule applies to episodes, and is tried in order to find out about whether there is a conical-hole in the well. If:

- The action being done just prior to the occurrence of the problem is drilling,
- The formation at the bottom of the hole is a hard formation,
- The type of bit used when the problem occurred is one of: jet-roller bit cv1-roller bit.

Then: There is suggestive evidence (.5) that there is a conical-hole in the well.

Here is a sample of an actual dialog between the drilling advisor and the drilling supervisor. User inputs follow the asterisks (* * *).

A Drilling Advisor Consultation:

(1) What is the name of WELL-159?
     ***AGF7-93E

(2) What is the profile of AGF7-93E?
     ***DEVIATED

the first formation of AGF7-93E will be referred to as:
                         ▪FORMATION-1▪

FORMATION-1:
     Please enter information about FORMATION-1

upper-limit meters	lower-limit meters	main-rock-type	homogeneous/ interbedded
(3)*** 747	806	SHALE	HOMOGENEOUS

▪ ▪ ▪

PROBLEM-1:
     Please enter information on PROBLEM-1

problem-type	prior-action	total-depth	casing-shoe depth
(13)*** STICKING	REAMING	1111 METERS	747 METERS

The first episode of the STICKING problem will be referred to AS:
                     EPISODE-1

EPISODE-1:
Please enter observations about EPISODE-1

	circulation type	type of injection pressure	rotation y/n
(14)***	NO-RETURN	OVER-PRESSURE— NON-STABILIZED	NO

(15) Please enter the composition of the drill-string starting from the bit (type *? for assistance):
***BIT•5/8 STAB9• 5/8 SHORTDC7•3/4STAB9•5/8. . .NDP5•

(16) What was the depth of bit at the time of the STICKING problem?
***1091 METERS

(17) What is the length of the SHORTDC section of the BHA (BHA.SECTION-3)?
***3 METERS

(18) What was the drilling method employed when the problem occurred?
***ROTARY

(19) Was reaming stopped just prior to the STICKING problem in order to add a single?
***NO

(20) What is the depth of the freepoint?
***UNKNOWN

(21) Is the SHALE formation interbedded with LIMESTONE between 806 meters and 960 meters a creeping formation?
***NO
• • •

• • •

The causes of sticking in EPISODE-1 are as follows:
BALLING-UP (.53).

The reason(s) that balling-up was concluded as the cause of sticking in EPISODE-1 are as follows:

The action at the time of sticking was REAMING. There are 4 stablizers in the BHA. There is evidence of balling-up material in the well. The upspeed of the cuttings is slow.

The list of possible curative treatments for EPISODE-1 are as follows:

You should work on the drill-string both upward and downward, and work on rotation.

Jar both upward and downward, using the jar currently in the BHA.

You may consider squeezing a lubricant and surfactant slug, because the drill-string is now motionless, and some of the non-spiralled elements in the BHA are in front of permeable formations.

Run the freepoint to determine the back-off depth, then (if possible) back-off and run a fishing jar able to work both upward and downward.

If you could not get the pipe free using one of the previous techniques, and depending on the results of the freepoint, you may consider using a cement plug and side-tracking.

The possible preventive treatments for EPISODE-1 are as follows:

If you could get the pipe free, you should now proceed to the following treatments during the next trip down:

Stop at different depths between 917 meters and 1111 meters to circulate complete bottom-up in order to clean the well.

In order to prevent any further problems of that kind, you may consider changing the mud characteristics as follows:

Add products for sticky shales.

You should change the BHA composition according to the following recommendations:

You should incorporate a safety joint in the BHA.

When drilling again, circulate with an increased flow rate around 1925.0 1/mn in order to have a good upspeed of cuttings.

Figure 22 reveals the secret behind our very rapid development of the drilling advisor. This project was completed in about 6 elapsed months. This speed results from two things: excellent knowledge engineering tools and experienced knowledge engineers. The tools used in this development are shown in the figure. These tools evolved from many previous domain-specific applications. The common properties of advisors of this sort have been abstracted into tools that make it easy to generate new domain-specific variants of the same generic form.

In summary, I would characterize the current state of technology in these ways. First, it is an extremely valuable technology and this is recognized by industry and government. Second, the technology has three primary limitations.

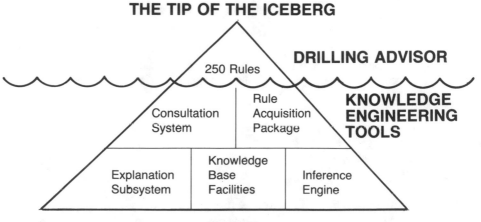

Figure 22.

- A shortage of skilled personnel, expecially senior project leaders
- Knowledge engineering tools are few and primitive
- The work is very difficult

As a consequence, demand for the technology currently surpasses the available supply. Only a few significant projects can be done in the near future.

I would like to turn now to the question of whether this technology is relevant to your business. To do that, I have drawn up a list of the most common reasons other businesses were led to exploring this technology. Figure 23 lists the most frequent situations that instigate such initiatives. Possibly, you may find one or more of these familiar.

On the other hand, knowledge engineering opens up whole new possibilities. To generate some new product idea, for example, simply apply the following heuristic rule: If there is a population of people who regularly buy knowledge or who could buy knowledge to avoid costly problems or who routinely perform tedious knowledge-based inference in their work, conjecture that a commensurate knowledge system product would appeal to that market. Figure 24 illustrates a very small number of the possible innovations.

This brings me to the end of my survey. Figure 25 reviews the major points of the paper. I have tried to give you an insight into where the field of knowledge engineering came from, what problems it solves today, and where the technology may be headed in the future. I see this field as a new industry based upon a break-through technology. But unlike some other promising technologies whose potential lies off in some remote future, this

---

SITUATIONS THAT INSTIGATE
KNOWLEDGE ENGINEERING INITIATIVES

- The organization requires more skilled people than it can recruit or retain.
- Problems arise that require almost innumerable possibilities to be considered.
- Job excellence requires a scope of knowledge exceeding reasonable demands on human training and continuing education.
- Problem solving requires several people because no single person has the needed expertise.
- The company's inability to apply its existing knowledge effectively now causes management to work around basic problems.

Figure 23.

KINDS OF PRODUCT INNOVATIONS NOW POSSIBLE

- Active instruments that solve problems by producing answers instead of data.
- Systems that explain how they work and how to use them.
- Corporate memories.
- Reasoning aids and thinking prosthetics.
- Accident-proof machines.
- Automated sellers.
- Hypothesis and expectation management systems.

Figure 24.

KNOWLEDGE ENGINEERING IN PERSPECTIVE

- The power is in the knowledge
- Mining, molding, assembling, and refining expert knowledge are the major tasks
- Familiarity with appropriate software tools can drastically reduce costs and speed development
- Managers must grow to harness this new technology: it's not business as usual
- This the **the** area of maximum growth potential for computer applications

---

Figure 25.

one offers rapid and high-value paybacks in the near-term for appropriate problems. This insures the continuous expansion of the market for knowledge systems and of the industrial infra-structure of knowledge engineering itself.

2. References

Duda, R. O. & Gaschnig, J. G. Knowledge-based expert systems coming of age. *Byte,* 1981, *6*(9), 238–278.

Feigenbaum, E. A. Themes and case studies of knowledge engineering. In D. Michie (Ed.), *Expert systems in the microelectronic age.* Edinburgh: Edinburgh University Press, 1979.

Hayes-Roth, F. Building expert systems. In S. Parker (Ed.), *The encyclopedia of electronics and computers.* New York: McGraw-Hill, 1984.

Hayes-Roth, F., Waterman, D. & Lenat, D. (Eds.) *Building expert systems.* Reading, MA.: Addison-Wesley, 1983.

Levine, J. Making it happen.. *Venture,* March 1983,14–16.

Shortliffe, F. H *Computer based medical consultations: MYCIN.* New York: Elsevier, 1976.

Natural Language Interfaces: What's Here, What's Coming, and Who Needs It

Madeleine Bates and Robert J. Bobrow
BBN LABORATORIES, INC.
10 MOULTON STREET
CAMBRIDGE, MA 02238

1. Introduction

The problem of getting computers to understand English[1] has been solved! Recent developments in artificial intelligence will soon make it possible for people to communicate with their computers in English! Natural language is the best way for people to interact with a computer! These and other claims abound in current popular literature and gossip. How true are they? This paper attempts to answer that question.

The term "natural language" (NL) is very deceptive. Everyone has an intuitive feel for what it means to communicate in natural language, but it is very difficult to make this notion precise. (The problem is well-illustrated by recent fierce debates about whether chimpanzees that have been taught some sign language are really using language.) For most of the history of the human race, the only entities using "natural" language have been human, so it is difficult to separate linguistic capabilities from other human capabilities such as memory, reasoning, problem solving, hypothesis formation, classification, planning, social awareness, and learning. On one hand, we do not want to require a computer to have all of these capabilities before we say that it can use language; on the other hand, without these capabilities any computer system will use language differently than human beings do, and thus is open to the charge that it is not really using language.

Without defining precisely what NL understanding is, most people

---

[1] In this paper we use the terms "natural language" and "English" as if they were synonymous.

would accept Woods' statement that "natural language assumes understanding on the listener's part, rather than mere decoding. It is characterized
by the use of such devices as pronominal reference, ellipsis, relative clause
modification, natural quantification, adjectival and adverbial modification
of concepts, and attention focusing transformations. It is a vehicle for conveying concepts such as change, location, time, causality, purpose, etc. in
natural ways. It also assumes that the system has a certain awareness of
discourse rules, enabling details to be omitted that can be easily inferred"
(Woods, 1977). This characterization absolutely excludes systems that
merely use English words to replace symbols in what would otherwise be
an "unnatural" language.

Human conversational partners share a lot of information, can model
one another's knowledge and capabilities, can process huge amounts of
information (even conflicting information), and can update all of these
structures in amazing detail as the conversation progresses. Computers
are a long way from having this very general, very powerful, very broad-
based language capability. Fortunately, a more limited language capability
will suffice for many applications, and humans can easily adapt at least
some aspects of their language based on their knowledge of their conversational partner. If too much adaptation is required, however, the communication becomes unnatural even if it is conducted in English. Thus we
must carefully distinguish between natural language communication, natural
communication, and user-friendly interfaces.

A critical obstacle to the use of many computational resources such
as Database Management Systems (DBMS) and decision support systems
(DSS) is the mismatch between the needs of end users and their ability to
communicate these needs to the computer. The development of graphical
interfaces such as spread-sheet systems, menu systems and the "electronic
desktop" are important steps towards improving the interface for a class
of stereotyped, semirepetitive tasks. For many tasks, however, greater flexibility and control are needed, and NL interfaces can provide this capability
to a wide range of users. Such interfaces would allow users who are unfamiliar with the technical characteristics of the underlying database management system to query databases using typed English input. The output
could be plain data or a graphical representation of the required data. In
addition, NL can be used to specify the input to a decision support system,
and to pose questions to such a system.

2. When Is English the Most Appropriate Interface Language?

In the rush to make computers more accessible, it is easy to be taken in
by the following false argument: Not everyone who wants to use a computer
can or will take the time to learn a special language for dealing with it;

everybody already knows English (or some other natural language); therefore, the only way to get everyone to use computers is to let them use English. This section shows the flaws in this argument, and sets the stage for examining NL interfaces in a more realistic way.

Natural language may be useful when the user of a system does not know the capabilities or limitations of the system, when s/he cannot or will not learn a formal interface language, when the underlying interface is not user-friendly (and hence would be awkward to use even if the user were prepared to get technical), or when the nature of the task to be performed is not well-specified.

Even under the conditions stated in the previous paragraph, natural language is *not* useful when the content of interactions is so limited that the brevity of an artificial language (such as a menu of choices) is desirable; systems with a sophisticated interactive graphic and menu interface (such as Xerox's Star and Apple's Lisa) can be operated without English and with very little training in the use of the interface. This use of icons is effective only because the users understand completely the conceptual model underlying the domain (opening files, sending messages, etc.) so that the level of detail that has to be conveyed by the icons and by the user's manipulation of them is very limited.

English is not useful when physical controls are appropriate — imagine driving a car or playing a video game using written or spoken English! Thus, in graphics-oriented situations such as laying out a slide for a presentation, or in computer aided design, the exact placement of the elements of an image is best done with some form of pointing device such as mouse, light-pen, tablet or touch-sensitive screen. In these cases, English may still play a role in initially specifying the images to be placed on the screen, if the set of possible stored images is large and not readily broken down in a way that would make single or multiple menu selections appropriate. English is not useful for object identification when the user can more easily point to something (as with a mouse or a touch-sensitive screen) than describe it.[2]

One intermediate position between formal interfaces and English is the use of tree-like menu systems in which each choice of an item from a menu causes the display of a new menu dependent on that choice. This is a good alternative to more complex interfaces, provided that the speed of the display is comfortable for use, that the amount of data to be presented in any menu is not too large to be visually processed easily, and that the user can identify the branch s/he wants to take at any point without having any knowledge of the structure of the tree at lower levels. It is difficult in

---

[2] However, pointing can be as ambiguous as English. Does a particular pointing action refer to "that line," "that triangle," "that region of the screen," "the object depicted by that triangle," or something else?

such a system to skim through the space of interactions, and for some applications it is difficult to find the desired information without skimming through the whole menu structure.

Another intermediate step between English and an "unnatural" artificial interface is query by example (QBE). This database interaction language lets the user specify data to be retrieved from or stored into a database by filling in a "table skeleton" that looks something like the answer table s/he wants to see (in the case of retrieval). For requests that involve finding those elements of a database that satisfy a simple set of constraints this is a very transparent query mechanism — the user specifies the desired form of the result and the system finds every piece of data of that form and prints specified fields. QBE is not limited to such simple queries, however, and is in fact somewhat more powerful than most other query languages for relational databases.

Although many people find QBE easy to understand and use, it is effectively just "syntactic sugar" disguising the fact that one is actually programming in a relational calculus. The limitations of this approach are that while it may be quite "user-friendly" and natural to use for simple retrieval requests, these human engineering features do not necessarily scale as the complexity of queries and interactions increases. In particular, the user must be aware of the exact structure of the database (the set of relations in the database or in a particular view of that database), and be able to perform certain symbolic manipulations in order to create queries in the form allowed by the system. Thus, a system like QBE is most useful when the typical interaction is simple, or when the personnel who use it are willing to learn a particular "programming technique" in addition to their other professional duties.

Even an application as apparently restricted as using a DBMS does not necessarily make the choice of interface easy, because of the many different kinds of tasks a user might want to perform. The most obvious DBMS task is to simply retrieve data ("What were our sales last year?"), but one might also want to format that data ("Graph last year's sales by month"), enter new data ("Set my department's projected sales for next month to $87,500"). query the system about its capabilities ("How far back do your sales figures go?"), place standing orders ("Don't show anyone's first name"), or do a myriad of other tasks. Different modes of interaction may be appropriate for different types of tasks; even if NL is the mode chosen, the type of language the user produces may vary greatly depending on the task being performed! The ambiguity of NL is often an advantage in retrieval tasks, but can be a serious problem when updating. A good example of this was given in Kaplan and Davidson (1981): if someone says "Change Brown's manager from Jones to Baker," does this mean that Brown is to be moved from the group managed by Jones to that managed by Baker, or that Jones is being replaced by Baker as the manager of the group that Brown is in? In order to interpret this command correctly, the part of the system that produces the interpretation must have access to the data in the database (to see, for example, whether Baker is currently a manager). Many NL systems attempt to complete their interpretation without such access, and consequently have difficulty with input of this type.

Even a system that has excellent NL capability within a small domain (such as accessing data about sales figures) may not be useful for users who have no idea of the limitations of that domain, or who want to perform tasks outside the scope of the system. Such users may want to ask questions about the system's capability such as "What can you tell me about personnel?", and the system may not be able to give any kind of coherent answer.

3. The State of the Art of Natural Language Interfaces

In an attempt to jump on the bandwagon of NL interfaces, some software producers simply take their current system interface and modify it slightly so that it uses English words and thus, at first glance, looks like it can understand English. One way to detect such exaggeration is to compare the "English" that is allowed with the underlying interface. If there is a fairly clear correspondence between the two, then very little NL processing is going on. Even if one does not have access to the underlying interface, it is usually easy to trip up such systems by giving them deliberately complex, ambiguous, or subtle language, using some of the constructions discussed in section 4.

Another distinction, and one that is harder to detect by simply observing the system in operation, is what Moore (1981) calls special-purpose vs general-purpose systems. General-purpose systems have the domain-dependent knowledge clearly separated from more general syntactic and/or semantic knowledge; such systems are of great interest to researchers. Special-purpose systems have knowledge about their particular application domain built in at very low levels of processing; they may, for example, be able to recognize units around a key word like "sales" but may not depend at all on general linguistic entities such as noun phrases. They may have special rules of inference for deducing new information from old, but the rules are formulated only for the particular application domain, not in general terms.

By mixing the domain model, database model, syntax and semantics of a particular domain, special-purpose systems[3] can achieve high per-

---

[3] Sometimes called *semantic grammar systems* or *pragmatic grammars* because they combine the semantics and/or pragmatics of the domain directly with syntactic analysis in a single grammar.

formance for that domain. Their drawback is that it is difficult or impossible for anyone but the original system designer to make significant changes to the system, and it must be almost entirely rewritten if a new domain is required. Experts such as Moore (1981) have said that it takes between 2 months and 5 years for programmers *experienced* in building these systems to produce a special-purpose NL front-end for a small but useful domain.

On the other hand, general-purpose systems offer the promise of easy transportability from one domain to another, by changing the lexicon and the domain-dependent semantics. Their disadvantage is the long development time required to produce the domain-*in*dependent components, and the fact that for *some* applications this approach brings more power to bear on the problem than is necessary, with a corresponding price tag. A critique of this approach is presented in Shwartz (1982). Research systems such as TEAM (Grosz, 1982, 1983; Grosz, 1982b) and IRUS (Bates & Bobrow, 1983) use this model, but are not yet commercially available. It will probably be several years before general-purpose (by this definition) systems begin to be widely available, but when they are, the effort required to adapt them to a particular application will be a few months (or even weeks).

There have been a lot of publications in the research literature about natural language interfaces. Most, but not all of them, focus on general-purpose systems. Some of these papers describe research systems that are being used to investigate various aspects of the NL problem, or are offered as "proof by example" that (limited) NL understanding is possible (Codd et al., 1978; Epstein & Walker, 1978; Ginsparg, 1983; Guida & Tasso, 1983; Hendrix, 1977, 1982; Hendrix et al., 1978; Shapiro & Kwasny, 1975; Spiegler, 1983; Thompson & Thompson, 1983; Walker & Hobbs, 1981; Waltz, 1977; Warren & Pereira, 1981). Others try to present general issues and problems relating to applied NL interfaces, particularly for database access: (Kaplan & Ferris, 1982; Moore, 1981, 1982; Templeton & Burger, 1983; Woods, 1977). Several conferences have had panels or sessions devoted to this subject (ACL, 1981, 1982, 1983), and several special issues of journals have also focused on it (Kaplan, 1982; Waltz, 1977). These research successes imply that the technological basis for commercial success has been achieved. Commercial ventures using this technology have recently begun to appear, and more are sure to follow.

The first major commercial NL interface system, INTELLECT, is produced by the Artificial Intelligence Corporation in Waltham, Massachusetts. It provides access to a number of DBMS on IBM, Honeywell and Prime computers. The initial versions of the system had to be tailored to the user's domain with several weeks of effort by the AI Corp. for each installation. More recently this tailoring has been done by systems staff in end user installations with the aid of documentation and training provided by

AIC. INTELLECT handles a comfortable subset of English, but can get caught on some complex or uncommon sentence structures. It can usually process telegraphic English, and has some spelling correction facilities. When the system cannot parse the input using its standard grammar, it falls back to an "error correction" strategy without warning, and can generate inappropriate translations. While the system gives some feedback as to how the user's query has been interpreted, it is not always easy to find out how to rephrase an incorrectly interpreted request to obtain a proper interpretation, nor even if that is possible. The system seems to have little or no model of various user domains, and thus leads to a mode of interaction which is often little more than an English restatement of a formal query, using the actual names of database fields. The end user must have a fairly detailed model of the database structure.

Another recent entrant into the commercialization of NL processing is Cognitive Systems, Inc., of New Haven, CT. This company is selling custom built NL interfaces, at a cost of several hundred thousand dollars. The advantages of this approach are: (a) a virtually turn-key system is delivered, eliminating the need for customization of the system by the client; and (b), because the method of language processing depends far more heavily on semantics than on syntax, the system can often process telegraphic or ungrammatical English. A spelling corrector is also provided. On the negative side, the system is not readily modified by the end user or system personnel at the user's installation.

One of the newest companies to enter this field is Symantec, of Sunnyvale, California. Their first product has not yet been announced, but it will be a DBMS interface for microcomputers that will be implemented in Pascal and will probably sell for under $500. The English capability will naturally be limited because of the small size of the target machines, but users will be able to build up their own vocabulary and syntax.

4. How Should Prospective Users Judge NL Systems?

In this section we present a number of topics that should be investigated when examining a system that claims to understand English. It is important to keep in mind that the right question to ask is *not* "Does system X have feature Y?" (because the answer will almost never be a clear yes or no), but rather "How much of feature Y does system X handle, and how important is it to the application I have in mind?". In a general-purpose system, the system developers should be able to describe the mechanisms used to handle these issues; a demonstration of their use in one domain is fairly good evidence of their applicability to another domain. In the case of special-purpose systems, evaluation is more difficult, since the techniques used

may be more ad-hoc, a demonstration that is impressive in one domain
may not be relevant to the kinds of problems that will arise in a different
application.

Coverage and habitability. These two properties of systems are related,
but not identical. "Coverage" is a characterization of the linguistic competence of a system, while "habitability" measures how quickly and comfortably a user can recognize and adapt to the system's limitations. The
coverage of a NL system may be categorized in a number of dimensions,
some of which are discussed below.

- *Lexical coverage.* How large a vocabulary does the system have?
  The overall size of the vocabulary is not as critical as the relevance
  of the vocabulary for the application domain, though the system
  should certainly cover all the "closed class" words of English such
  as prepositions, conjunctions, articles, etc. In addition, since it is
  impossible for any system to have complete coverage, it is important
  to know how easy or difficult it is to extend the vocabulary of the
  system. What knowledge of linguistics and the internal structure of
  the dictionary is required? Can an end user add new vocabulary, or
  does it take an applications programmer with some short training,
  or must vocabulary always be added by the system developers? It
  is also important to distinguish between new words that are essentially synonyms for existing words, and new words that involve new
  concepts for the system.
- *Syntactic coverage.* What is the range of syntactic phenomena which
  the system can deal with? Does the system handle complex verb
  forms, relative clauses, various question forms, passives, compar-
  atives, subordinate clauses, time and place adverbials, measure
  expressions, ellipsis, pronominalization and conjunction? While this
  is the most well studied aspect of natural language understanding,
  there is not as yet a benchmark against which to test a system, nor
  even a generally agreed upon list of phenomena. The closest thing
  to this is the list of phenomena given by Winograd in his recent
  book on syntactic processing (1982).
- *Semantic coverage.* How much does the system *understand* about
  the domain? For a DBMS retrieval system, does the system have a
  model of the semantics of the applications domain, or does it merely
  make a direct translation of certain English phrases into specific
  queries in a formal retrieval language? This is particularly important
  if the system is to be able to access new databases, or to work when
  old databases are restructured. There is clearly a major difference
  between having to ask "Is there an employment record for Jones
  with Acme Co. in the employer field?" and "Did Jones ever work for

Acme?". If the system treats the latter question as simply a variant of the first, then it will not be able to handle such a query if the database is modified to list the employees for each company (but not the companies for each employee), nor would one expect it to be able to handle "Did Jones ever work for division 5?" or "Did Jones ever work for Smith?".

While a system with extremely large coverage is likely to be habitable, even systems with very limited coverage can be habitable if properly designed, and systems with wide variations of coverage may be less habitable than ones with uniformly smaller coverage. The critical issues are whether the system has enough coverage to let users meet a reasonable proportion of their needs (i.e., is there at least one way to express everything a user really needs to say), whether the user can quickly find the appropriate way of expressing a request, and whether the user can easily learn to avoid the system's blind spots.

As examples of two ways in which even a system with reasonable coverage may have reduced habitability, we point out pitfalls involving language output and semantic grammars. In both cases, the problem is similar — the user is led to believe that the system has capabilities which are beyond it, and there is no clear indication of the boundaries.

In many applications, NL output is not necessary, but almost certainly some English will be presented to the user, even if it is only canned text. English output from a computer system will either be prestored strings or generated text (Mann et al. 1982) that comes from a different knowledge base than that used by the language understanding part of the system, instead of being integrated as in humans. This means that the language that can be expressed by a computer system exceeds its comprehension, a situation which is precisely opposite that of humans! Human users of a system will, very naturally and unconsciously, be influenced by the computer's language, and will assume that the computer can understand, at a minimum, the kind of language it produces. Thus, a good goal would be to ensure that the vocabulary in the output is understood by the interface, and that the syntactic constructions used are within its syntactic coverage. Since building a natural language understanding system with broad lexical, syntactic and semantic coverage is a substantial undertaking, it is likely that the goal of matching input capability with the system's output will have to be modified. In this case, the user should be clearly informed of the way in which the two capabilities differ. Finally, even if the two capabilities are matched, there is another possible pitfall. In normal conversations people typically use pronouns and other anaphoric expressions like "that purchase order," "those salespeople" and "the average" to refer to entities introduced into the conversation by their dialog partner. If the system uses canned text for output, or even if it synthesizes English output as needed, it will not be able to understand such anaphoric expressions unless it maintains a model for everything *it* has said.

The difficulties in achieving habitability with a semantic grammar are based on the fact that without great care such grammars can give users misleading clues as to coverage. If the system can understand both "list the salespeople who have been under quota for two months," and "what salespeople have been under quota for two months," and the system can understand "list the products that Jones sold to Acme," then the user might reasonably expect the system to understand "what products did Jones sell to Acme?". In a semantic grammar, however, the system has different portions of the grammar for each verb or closely related verb, and it is easy to allow for some verbs to be used in both relative clauses like " . . . which were V'ed by Jones" and in WH questions, and not allow other verbs in *both* constructions.

Inference. This is the art of drawing logical conclusions based on the data in the database and general knowledge of the subject domain. It is often the case that retrieving only data that is explicitly stored in a database is insufficient to meet a normal user's needs. Users will assume that the system has the ability to infer new information from that already in the database. (This is particularly true if the user does not have detailed knowledge of the database.)

As an example of a simple inference (sometimes called the "navigation problem"), suppose a database contains records about employees and records about jobs the company has performed for clients. The employee record has a field for jobs the employee has worked on, and the job record has a field for the client's name. Someone accessing this database might naturally ask "Has Ellen Matthews ever worked for Adams Co.?" Notice that in order to interpret this question correctly, the system must be able to follow the chain of reasoning that Matthews has worked for Adams if she has worked on a job that had Adams as the client, but no job was explicitly mentioned in the query and no relation 'work for' exists in the database.

It is very important to understand the *level* of questions or commands one can give to a NL interface. Will the user be able to ask a high-level question that requires more processing than simple information retrieval ("Is the art department in financial trouble?") or must s/he ask a number of simple questions and then integrate the results ("Show me the art department's projected and actual expenses for the last 6 months. List the art department's projected expenses and work schedule for the next 3 months.")?

End user control of interpretation. Suppose you want to ask "What is the largest division in the company?" You could mean largest in terms of number of employees, number of employees of a particular type, gross sales, or some other metric. Either the system has some built-in metric or it doesn't. If it does, it may or may not be what you meant. If it isn't what you meant, how do you find this out (the answer "Division 4" probably won't help) and can you change it? If the system doesn't have a default metric, it might have a set of metrics that it can ask you about, but you don't want to see the question "Do you mean largest number of employees or largest building area or highest sales?" every time you use the term large. Ideally, the user should be able to set temporary (or permanent) "standing orders" that will be interpreted in context, but this is currently possible only in a limited way.

Use of pronouns. Any NL system will claim that it can handle pronouns (he, her, it, they, their, himself, etc.), because they are so widely used in English, but every system will have limitations in this regard because pronoun use can be extremely complex. For example, pronouns usually refer to objects explicitly mentioned in previous discourse, but sometimes they can refer to objects mentioned later ("After he transferred from Department 22, did John Jones work in Division 6?"). Pronouns can also refer to actions ("Did Smith ever come to work later than 10am? How often has he done that?"). In NL interfaces users find it perfectly natural to use pronouns to refer to objects in the computer's previous response, not just objects in their own language (Q: "How many projects are ahead of schedule?" A: "Project 356–3" Q: "Who is in charge of it?").

Other kinds of reference. Pronouns are a specific case of a linguistic phenomenon called anaphoric reference, in which one refers to things without using their full names. Even things that have not been mentioned explicitly can be referred to, if it is "obvious" that they should be inferred from the previous context. For example, the multi-sentence utterance "Seven contracts were concluded last month. Those profits will set a new record." uses the phrase "those profits" to refer to the profits of the contracts just mentioned.

Every "natural" DBMS interface must provide some ability to specify items on the basis of previously computed aggregates, for example "products whose sales are at least 80% of the average sales of the ten most profitable products."

Ellipsis. In conversation, people often leave out large portions of sentences, assuming that the missing parts can be filled in by the listener who shares the context being discussed. For example, a user might want to make the following series of queries: "How many people did we hire last month?", "The month before?" "How many do we expect to hire next month?"). It is easy to be fooled into thinking that just because a system handles a few examples, it can handle any kind of ellipsis.

Quantification. The use of words like "some," "every," "all," and "any" can complicate NL understanding, because their interpretation often depends on wide-ranging "common sense" knowledge, or on detailed knowledge of the particular domain. The queries "Did every person in department 5 submit their trip report?" and "Did every person in department 5 consult their department manager?" are structurally equivalent, but the first case refers to multiple trip reports and the second case to a single manager.

Negation. Negation is particularly tricky when combined with quantification. Does "All of the projects weren't completed on time" mean that none of the projects were completed on time, or that some were and some weren't?

Time and Tense. This is currently an open research issue. There are no general mechanisms for effectively and efficiently representing events and objects that change over time. Fortunately many database applications do not have to be concerned with this issue, since they contain historical data which might be updated or corrected but which does not contain complex time relations.

Conjunction and disjunction. And's and or's are extremely common in English. Often they are used to join complete units ("the book and the author") but sometimes they can be used to join discontinuous segments ("I adjusted for and calculated next quarter's overhead"). Handling simple conjunctions is within reach of current systems, but the combinations of conjunction with ellipsis and other phenomena is still an open problem in computational linguistics.

Telegraphic input. Although full English sentences are easy to say, people who have to type a lot frequently want to abbreviate their input by dropping out "unnecessary" words. For example, "Show sales last year midwest by salesman" is easily understood (by humans) as a paraphrase of "Show (me) (the) sales (from) last year (in the) midwest, (graphed with sales) by salesman." Of course, in the appropriate context it might also mean "Show (to the sales department) (the figures from) last year (graphed with the) midwest (sales) by salesman." One important point to remember about this capability is that, although it is desirable, one pays for having it by increased potential for misunderstanding and decreased ability to use the finer points of grammatical structure to influence the processing of full as well as telegraphic input.

Ungrammatical input. Closely related to the notion of telegraphic input is that of ungrammatical input. In fact, some of the same techniques can be used to handle both kinds of nonstandard language.

All of the issues itemized above represent problems that have been at least partially solved in general-purpose research systems. Here we present some highly desirable attributes of systems for database retrieval that are not so well-understood in general terms but may be available in limited form for particular applications in the very near future.

- *What-if capability.* This is nearly essential in DSS systems. Simple specifications of conditions are easy to handle, but complex specifications present serious problems, particularly if they are expressed incrementally and modified during a dialogue.
- *Presentation of output.* This includes formatting reports and tables, interfacing to graphics modules, and generating English output. Simple capabilities are available now, and will rapidly expand.
- *Tools for altering the domain and file structures.* Systems that have a very direct correspondence between the input language and the retrieval language can be modified by end users (or systems programmers at the end user's organization), but more sophisticated NL capability implies the need for most customization to be done by the developer of the NL system. Software tools that will make it easier to develop, expand, and modify domain-dependent information and DBMS-dependent information will only gradually be developed.
- *Implementation in a work station.* For many applications, it is undesirable to use mainframe computer resources to process English queries and commands. Soon some NL systems will fit comfortably in individual work stations or personal computers, and will be able to locally translate the user's input into a sequence of commands to be sent to the DBMS on another machine.

Perhaps a word is in order for people in moderate-sized organizations with considerable in-house expertise in database systems (or DSS, or whatever) who are thinking of developing their own NL front-end from scratch. There are many more complex and subtle problems in NL processing than this paper hints at, and for a nonexpert to attempt to develop an adequate NL interface is to invite disaster. At best, one could get a system that is adequate for limited purposes but is difficult or impossible to expand and maintain; at worst, the expenditure of a lot of resources may fail to produce a useful system at all.

There are (at least) two approaches to looking for NL capability in a computer system. One is to look at available systems and say "if I had it, what could I do with it?" This is likely to be misleading, since it is very easy to infer from a few examples that the system can do more than it actually can. A better approach is to determine *in advance* what kinds of interactions one would like to be able to have with the machine (perhaps by taking protocols of a simulation, as in Bates & Sidner (1983), or just by asking potential end users of the system to write down a few dozen examples). Armed with this unbiased language sample, one can ask "Will system X be able to handle this input?"

5. Conclusion — The Future

In the next few years we can expect to find natural language interfaces to a wide variety of computer systems, including database systems, graphics packages and decision support systems. This already large market is certain to grow as personal work stations and network access to data and decision support services become widely available.

Some organizations will choose to develop their own NL interfaces in-house, others will buy that capability elsewhere. Because the development of language systems requires a much different programming approach than, say, accounting or database packages, the in-house systems will be almost entirely special-purpose and difficult to modify as the needs of the people using them grow. There will be a large number of companies offering to build special-purpose systems on a contract basis, and fewer offering general-purpose systems (because of the very limited supply of experts needed to develop them, and the lengthy development cycle).

The subject areas for NL applications will be very broad — systems for inventory control, purchasing, contracts management, computer resource allocation, etc. can be built with modest effort. Some vendors will aim for one or more well-defined user communities and develop specialized packages, others will build a family of more general, tailorable systems. It will not be easy for the purchaser of a NL system to judge whether the system is capable of meeting the demands of the proposed application. This problem will continue to require expert advice and consulting.

The point we have tried to make is that the technology for useful, cost-effective, natural language interfaces *is* available now and will begin to have a major impact on database retrieval and other areas in the very near future. However, these interfaces will not behave like a human conversational partner, so users must carefully examine such systems to understand their capabilities and limitations.

6. References

Association for Computational Linguistics. *Proceedings of the 19th Annual Meeting of the Association for Computational Linguistics,* ACL, Stanford University, 1981.

Association for Computational Linguistics. *Proceedings of the 20th Annual Meeting of the Association for Computational Linguistics,* ACL, Toronto, Ontario, Canada, 1982.

Association for Computational Linguistics and the Naval Research Laboratory. *Conference on Applied Natural Language Processing,* ACL, Santa Monica, CA, 1983.

Bates, M., & Bobrow, R. J. A transportable natural language interface for information retrieval. In *Proceedings of the Sixth Annual International ACM SIGIR Conference,* June 1983.

Bates, M. & Sidner, C. L. A case study of a method for determining the necessary characteristics of a natural language interface. In P. Degano and E. Sandewall (Eds.), *Integrated Interactive Computing Systems.* New York: North Holland Publishing Company, 1983.

Codd, E. F., Arnold, R. S., Cadiou, J M., Chang, C.L., & Roussopoulis, N. *RENDEZVOUS Version 1: An experimental english-language query formulation system for casual users or relational data bases* (Tech. Rep. RJ2144). San Jose, CA: IBM Research, January 1978.

Epstein, M. & Walker, D. Natural language access to a melanoma database. In *Proceedings of the Second Annual Symposium on Computer Applications in Medical Care,* Washington, D.C., November 1978. Also SRI Technical Note 171, September 1978.

Ginsparg, J. M. A robust portable natural language database interface. In *Proceedings of the Conference on Applied Natural Language Processing,* February 1983.

Grishman, R., Hirshman, L., & F., C. Isolating domain dependencies in natural language interfaces. In *Proceedings of the Conference on Applied Natural Language Processing,* February 1983.

Grosz, B. *TEAM: A Transportable natural-language system* (Tech. Rep. No. 263). Menlo Park, CA: SRI Artificial Intelligence Center, April 1982.

Grosz, B. J. Transportable natural-language interfaces: Problems and techniques. In *Proceedings of the 20th Annual Meeting of the Association for Computational Linguistics,* Toronto, 1982.

Grosz, B. J. TEAM, a transportable natural language interface system. In *Proceedings of the Conference on Applied Natural Language Processing,* February 1983.

Guida, G., & Tasso, C. IR-NLI: An expert natural language interface to online databases. In *Proceedings of the Conference on Applied Natural Language Processing,* February 1983.

Hendrix, G. G. *The LIFER manual: A guide to building practical natural language interfaces* (Tech. Rep. Technical Note 138). Menlo Park, CA: SRI International, February 1977.

Hendrix, G. G. Natural-language interface. *American Journal of Computational Linguistics* April–June 1982, *8*(2) 56–61.

Hendrix, G., Sacerdoti, E., Sagalowicz, D., & Slocum, J. Developing a natural language interface to complex data. *ACM Transations on Database Systems* June 1978, *3*(2), 105–147.

Kaplan, S. J. Special section — natural language. *SIGART Newsletter,* January 1982, 94–95.

Kaplan, S. J., & Davidson, J. Interpreting natural language database updates. In *Proceedings of the 19th Annual Meeting of the ACL.* Association for Computational Linguistics, Stanford University, June 1981.

Kaplan, S. J., & Ferris, D. Natural language in the DP world. *Datamation,* August 1982, 114–120.

Mann, W., Bates, M., Grosz, B., McDonald, D., McKeown, K. & Swartout, W. Text generation. *Americal Journal of Computational Linguistics* April–June 1982, *8*(2), 62 69.

Moore, R. C. *Practical natural-language processing by computer* (Tech. Rep. Tech. Note 251). Menlo Park, CA: SRI International, October 1981.

Moore, R. C. Natural-language access to databases — Theoretical/technical issues. In *Proceedings of the 20th Annual Meeting of the Association of Computational Linguistics,* Toronto, 1982.

Shwartz, S. P. Problems with domain-independent natural language database access systems. In *Proceedings of the 20th Annual Meeting of the ACL.* Toronto, June 1982.

Shapiro, S. C. & Kwasny, S.C. Interactive consulting via natural language. *Communication of the ACM, 18*(8), 1975, 459–462.

Spiegler, I. Modelling man-machine interface in a database environment. *International Journal of Man-Machine Studies, 1983, 18,* 55–70.

Templeton, M., & Burger, J. Problems in natural language interface to DBMS with examples from EUFID. In *Proceedings of the Conference on Applied Natural Language Processing.* February 1983.

Thompson, B. H. and Thompson, F. B. Introducing ASK, a Simple Knowledge System. In *Proceedings of the Conference on Applied Natural Language Processing.* February 1983.

Walker, D. E., & Hobbs, J. R. *Natural language access to medical text.* (Tech. Rep. Technical Note 240). Menlo Park, CA: SRI International, March 1981.

Waltz, D. L. Natural language interfaces. *SIGART Newsletter,* February 1977, (61), 16–17.

Waltz, D. L. An english language question answering system for a large relational database. *Communications of the ACM,* July 1978, *21*(7), 526–534.

Warren, D. H. D., & Pereira, F. C. N. *An efficient easily adaptable system for interpreting natural language queries* (Tech. Rep. 155). University of Edinburgh, February 1981.

Winograd, T. *Language as a cognitive process, Volume I: Syntax.* New York: Addison-Wesley, 1982.

Woods, W. A. A personal view of natural language understanding. *SIGART Newsletter,* February 1977, (61), 17–20.

Natural Language Communication with Machines: An Ongoing Goal*

W. A. Woods

BOLT BERANEK AND NEWMAN INC.
10 MOULTON STREET
CAMBRIDGE, MA 02238

This paper is concerned with issues of man-machine interaction in decision support systems for high level decision makers. It discusses components that such systems should have, what the current state of the art is with respect to such systems, and how current research in artificial intelligence is leading toward solving the remaining problems. Topics covered include natural language syntax and semantics, models of the beliefs and goals of the user, and knowledge-based helpful systems.

1. Introduction

Suppose that you had five years in which to design a really good decision support system for high-level decision makers. Where would you start? What would you try for? How would you do it? I will try to give a sketch of the components that I think such a system should have, what the current state of the art is with respect to such systems, and how current research in artificial intelligence is leading toward solving the remaining problems.

I should begin by saying that what I have in mind by decision support is not a package of statistical decision procedures with respect to whose framework the decision maker is to express his options and valuations, after which the system will determine the optimal decision. Rather, I am concerned with the situations in which the decision maker has a problem,

* The research described here is the result of a large group effort involving several projects over a number of years. Participants in the group have included Madeleine Bates, Robert Bobrow, Ron Brachman, Phil Cohen, Brad Goodman, David Israel, James Schmolze, Candace Sidner, Marc Vilain, and Bonnie Webber. This research has been supported by ARPA under contract N00014–77–C–0378, monitored by ONR, and by ONR under contract N00014–77–C–0371.

has not yet determined his options, much less his valuations, and is instead trying to come to understand the nature of the problem. Assuming that there is a computer system that contains extensive data among which the relevant information toward characterizing the problem may lie, how will the decision maker find the relevant information and discover the patterns of information which will help him to understand what the problem is? How can a computerized system facilitate this task?

I am concerned with the unanticipated, nonstandard decision situations for which one cannot expect to have a predetermined package that displays just the right information in the right way to give the decision maker what he needs. That is, for the situations (which I suspect are quite frequent) in which regular monthly or weekly reports do not give the whole picture, but rather suggest questions that require further investigation, how do you construct a system that truly facilitates the investigative digging that is required to discover what is going on? How do you help a manager discover what is wrong when some aspect of the business is not going as expected?

A principal objective is to make the system sufficiently flexible that the decision maker can get information presented in whatever manner he finds helps him understand the situation, and to make it sufficiently intelligent and fluent that he can do this without having to take his attention away from the problem he is trying to solve and devote it instead to the issue of how to get the computer to do what he wants.

I will begin by describing a project at Bolt, Beranek and Newman that is addressing these issues. This project is attempting to make simultaneous, coordinated advances in a number of fundamental areas necessary for improving human communication with computers. These include fundamental techniques for the representation of conceptual knowledge, techniques for constructing helpful systems that can reason about a user's plans and goals, and techniques for efficiently parsing natural language and performing knowledge-based inference.

2. Knowledge-Based Language Understanding for Decision Support

BBN's research in Knowledge Representation and Natural Language Understanding is aimed at developing techniques for computer assistance to a decision maker in understanding a complex system or situation using natural language control of an intelligent graphics display. The motivating need is that of a military commander in a command and control context — especially in crisis situations. In such situations, not only does the commander need certain information in order to make his decisions effectively, but in complex situations, this requires the presentation of that information in a form that is matched to the abilities of human comprehension. A hypothetical scenario to illustrate the kinds of interaction we envisage is given below:

1. Cdr: Show me a display of the eastern Mediterranean.

    [Computer produces display.]

2. Cdr: Focus in more on Israel, and Jordan.
    [computer does so]

3. Cdr   Not that much; I want to be able to see Port Said and the island of Cyprus.

    [computer changes scale and window to include the desired features]

4. Cdr: Now show me the positions of all U.S. and Soviet vessels in the area.

    [computer does so, and makes a default assumption for displaying the difference between U.S. and Soviet vessels]

5. Cdr: Where is the John F. Kennedy?

Computer: Two hundred miles to the west of the point displayed.

    [The ship is not on the screen, so the system displays a point at the left edge of the display.]

6. Cdr: Show me the course tracks for the Soviet vessels for the last five hours.

    [Computer does so.]

7. Cdr: What kind of ship is that?

    [Points to a Soviet vessel.]

Computer: Soviet missile cruiser.

8. Cdr: Show me the other missile cruisers, and display all vessel types with two digit code.

    [Computer blinks or flashes all of the missile cruisers for 2½ seconds and displays with each vessel the two digit code (assumed previously agreed on by the commander).]

9. Cdr: Remove the course tracks, and show small dots with one-hour course tracks for any known Soviet aircraft in the area.

    [Computer does so.]

    [Commander makes his assessment of the situation and makes appropriate orders for his forces.]

10. Cdr: Remove the planes and track the Soviet vessels for the next four hours. Show any deviations from current course double intensity and ring bell when detecting course change. Flash vessel changing course for 10 seconds.

    [Computer accepts standing orders for continual monitoring and conditional future behavior.]

Notice that the user's utterances include imperatives to be taken as direct commands, as well as declaratives to be taken as indirect commands (as in exchange 3). Notice also that the system is given considerable latitude to plan its response to match what it "thinks" the user expects, rather than being meticulously instructed at a detailed level (for example, in choosing the exact boundaries of the region to display in response to utterance 1). In exchange 3, the user has simply stated the objectives that he wants to achieve at that point and left it to the system to determine how to achieve them.

We have conducted an experiment in collecting protocols of users interacting with simulated versions of the kinds of system we envision (Sidner, 1982). Our analysis of those protocols has convinced us that the behavior exhibited in exchange 3 above is the "tip of the iceberg" of a much more varied and common linguistic use. In particular, people often discuss a wide variety of changes they require in a system's response either because they have changed their minds or because what they said the first time didn't do what they wanted.

To explore the problems implicit in this scenario (which has not itself been implemented) and to develop techniques for dealing with them, we have been developing experimental prototypes in simplified domains. One of these is a system to support inspection and debugging of ATN grammars (Brachman et al., 1979). This system parses and interprets English requests, synchronized with pointing events on a screen, and produces appropriate display actions on a bit map graphics display in response. It permits a user to request portions of a display to be shown, objects in the display to be made visible or invisible, attributes of objects pointed to to be displayed, and screen windowing to be changed by means of statements of constraints on what is to be visible.

The system includes a sophisticated knowledge representation system, a comprehensive grammar of English, powerful general tools for natural language processing, and experimental capabilities for tracking the focus of attention in an ongoing communicative dialog, modeling the beliefs and goals of the user, recognizing the plan that underlies the user's utterances, and planning helpful responses to the perceived goals of the user.

A major accomplishment in this work has been the development of the knowledge representation system KL–ONE (Brachman et al., 1979) and its use in the construction of the experimental prototype. KL–ONE is used to organize the semantic interpretation rules used to interpret sentences, to organize the models of the user's goals and beliefs (which are used to fill in details that are not explicit in the input), and to organize the knowledge of displays and display forms that are used to draw the pictures on the screen. The knowledge structuring capabilities of KL–ONE have proven themselves very powerful in this system, and the extent to which the same structures have proven useful in qualitatively different parts of the system gives evidence of the robustness of these capabilities.

Although the major thrust of our work is to address fundamental issues in a theoretically sound and general way (attempting to avoid the pitfalls of optimizing on aspects of particular applications), the results of the work to date have included not only increased understanding of the fundamental issues but also concrete subsystems that have been found useful by other groups in the scientific community.

3. *The Need for Fluency and Conceptual Power*

The underlying assumption of the BBN project is that in a crisis situation the commander needs an extremely flexible system, capable of manipulating large amounts of data and presenting it on a graphical display in a variety of ways until he feels satisfied that he has a grasp of the situation. Such a system would have abilities to display many kinds of map overlays, an ability to change the kinds and amounts of detail shown, an ability to conveniently construct unique displays to suit the situation at hand, as well as the ability to display tabular and graphical information and present textual material in ways that are easily comprehensible. This situation is not fundamentally different from the needs of a manager with a complex business decision to make.

In such circumstances, the display that the user wants and the modifications to it that he will subsequently want must be described in a highly fluent and expressive language, at a level of abstraction appropriate to the user's intent. One must not require the equivalent of a graphics programmer in order to obtain the displays required. Rather, one needs a system that is able to accept an abstract specification of the essential details of what should be in a display, and then intelligently and effectively determine the remaining details necessary to actually produce that display. This is true whether the actual specification of requests to the computer system is done by the decision-maker himself or by one or more subordinate specialists.

If the language of such a system is to be matched well to human cognitive abilities, it appears necessary for it to include a number of aspects of ordinary natural language, such as the use of pronouns, the ability to take an incomplete specification and fill in the details on the basis of prior knowledge, and the ability to take a specification that would be potentially ambiguous out of context and determine the intended meaning. While artificial languages could perhaps be designed with the necessary properties, it is not obvious that one could do better than English as a language with sufficient power for expressing all of the needs of a manager in a complex decision task, while retaining the naturalness of use of English. Moreover, if one succeeded, it is likely that most of the computer processing difficulties inherent in understanding English would be present for this artificial language as well.

4. The Rationale for Natural Language Understanding

There are many advantages of natural language as a communication channel between a man and a machine. One of them is that the man already knows the natural language, so that he does not have to learn an artificial language nor bear the burden or remembering its conventions over periods of disuse. It also avoids his consciously translating (programming) his requests into the artificial language from the form in which they occur to him (presumably in a form very close to natural language). Expecially for high-level personnel who use a computer system infrequently, or at least do not spend a major portion of their time dealing with the machine, these extra burdens of artificial language impose a severe barrier to the use of a machine.

Even for technical specialists who deal with a computer constantly, there is a distinction between the things that they do often and remember well, and many other things that may require consulting a manual and/or much conscious thought in order to determine the correct machine "incantation" to achieve the desired effect. For naive, inexperienced users, almost every transaction with the machine is of this form and the difficulty of deciding how to express a request is even more severe.

Whether a user is experienced or naive, and whether he is a frequent or occasional user, there arise occasions where he knows what he wants the machine to do and can express it in natural language, but does not know exactly how to express it to the machine. A facility for machine understanding of natural language can greatly facilitate the efficiency of expression in such situations — both in speed and convenience, and in decreased likelihood of error.

A more important motivation for natural language understanding is the way that the underlying conceptual structures of English can match the user's conceptualization of the problem. Although most current natural language understanding systems do not achieve this goal, the understanding of the underlying English conceptual structure is far more important than the superficial resemblance to English syntax.

The problem of representation and use of conceptual knowledge in computers is of critical importance in a wide variety of applications. These include not only intelligent, knowledge-based systems, but also general programming languages and systems. There is growing evidence that effective use of computers by both novices and experts, and for both command interpreters and software development systems, is greatly facilitated by structuring the computer program to use conceptual structures that correspond naturally to the conceptual structures that people use to organize the same information.

The issue of understanding the conceptual structure underlying natural language becomes especially important when the data to be manipulated by the machine is fundamentally natural language data. Such situations occur with interoffice memos, computer mail, requisitions, procurement specifications, etc. Current data management systems deal with data that can be fit into a relatively small number of predetermined formats, and they support requests of a relatively straightforward class. Forcing data into such formats usually leaves many things unexpressible in the data base, and the artificiality of the resulting data structure often makes the expression of many kinds of requests either impossible, or a difficult programming task.

It is my belief that an effective communication system for man-machine interaction in complex decision-making tasks must be essentially a natural language system. What is essential here is not the use of natural English words, although that has considerable mnemonic value, nor necessarily natural English constructions, although that considerably eases the learning task and the processing load required to use the system, but rather the natural English conceptual structure normally used for communication by humans. This conceptual structure has evolved through centuries of trial and error to become a very effective means of communication for an open ended set of complex ideas, situations, and goals. It is not only unlikely that we could design an artificial structure to match it, but such an artificial structure would also be difficult to learn and use.

Note that natural language does not preclude the introduction of abbreviations and telegraphic shorthands for complex or high frequency concepts — the ability of natural English to accommodate such abbreviations is one of its strengths. Indeed, it is in the development of concise ways of saying relatively complicated things that natural language excels, and it is the development of a rich inventory of concepts in terms of which to express requests that make such a system effective.

5. Beyond Syntax and Semantics

Natural language communication assumes a certain level of understanding (rather than mere decoding) on the listener's part. It is characterized by the use of such devices as pronominal reference, ellipsis, relative clause modification, natural quantification, adjectival and adverbial modification of concepts, and various attention focusing transformations. It is a vehicle for conveying concepts such as change, location, time, causality, purpose, etc. in natural ways. It also assumes that the understanding system has a certain awareness of discourse rules, enabling details to be omitted that can be easily inferred. It is the presence of such capabilities that distinguishes what one should properly call a natural language understanding capability.

If the dialogues that I and others have considered can be taken as a fair sample, minimal requirements for a natural language system for use in decision-making applications include a facility for expressing quantification of actions or tests over sets of objects, for using adjectival and relative clause modification, for determining objects and sets of objects, for adverbial modification of verbs, for pronominal reference and definite noun phrase reference to objects or sets introduced previously in the discourse, for within-sentence pronominal reference, for conjunction and negation, for time and tense, and for extensive paraphrase variation in referring to objects and actions. In addition, it is highly desirable that the system be capable of understanding the various kinds of surface word-order transformations which people routinely apply (sometimes to eliminate ambiguity and sometimes to bring important aspects into focus), and that it be capable of dealing with various kinds of ellipsis and vagueness.

Most of these areas have been studied by computational linguists and linguists, and many of them are sufficiently well understood that techniques for handling them are fairly well in hand. There is, however, a frequent tendency for claimed "solutions" to a given phenomenon to handle in truth only a restricted subset of the phenomenon. For some tasks, the restricted-case solutions are sufficient for useful application, but in general the details of a proposed solution need to be considered carefully to determine whether it is adequate for a particular application. Thus, one must be wary of assuming that a system that claims to handle pronominal reference and ellipsis (say) will actually handle either general examples of these phenomena or ones that actually arise in a particular application (even in the application for which the system was designed).

6. The Need for Intelligent, Helpful Systems

Much of the time in communication with the system, the user will not say literally exactly what he means, and there are good reasons not to require him to do so. The major reason is that it is cognitively inefficient to be meticulously literal in one's communication (that's why computer programming is a time consuming and expensive activity). One of the major activities in programming a computer to do a complex task is the systematic specification of all of the details that would be left unsaid if one were instructing a human to carry out the same task. In the complex decision-aiding situations that we are considering, we cannot afford to require this degree of literal specification of detail. Rather, the system must know enough about the objectives of the user that it can fill in details in reasonable ways, asking the user for clarification occasionally, but only when absolutely necessary.

Moreover, the system should be able to use its general knowledge and the knowledge in its database to go beyond merely doing what was requested, to provide additional information that can be inferred to be relevant to the user's goal and not otherwise known to the user. For example, when a military commander asks how many of his interdiction fighters are equipped with a particular kind of radar during a mission planning operation, the system should volunteer information about how many of those radars are out of commission (unless it knows that the commander already knows that). That is, the system should go beyond the passive execution of the user's commands to infer the goal structure underlying those commands where possible, and to volunteer additional relevant information (usually in accordance to standing instructions as to what kinds of additional information should be offered in what situations).

Thus, in addition to understanding the syntax and semantics of language, a helpful decision system must share with the user:

1. Knowledge of the domain of discussion.
2. Knowledge of some of the user's intentions and goals.
3. Knowledge of what the user thinks the system can do.

This knowledge will make it possible for the user to interact with a flexible system that interprets his needs appropriately. For example, the user should be able to perform the following:

1. Request an action by the system or an effect to be accomplished where the level of description in the request is abstract and details are filled in by the system.
2. Ask questions whose proper interpretation depends explicitly or implicitly on the system's ability to infer some of the user's intentions.
3. Propose modifications of previous requests or of system responses where the system is to infer the relationship between the modification and the previous discourse.
4. Ask for clarifications, and then modify a request, where the system provides help in response to the request for clarification and properly responds to the modified request.

5. Order the system to modify its overall future behavior, where the system responds by changing its internal model of future action to conform with the order.

These features describe the kind of helpful system which I believe will be needed for complex decision-aiding tasks.

7. The Need to Recognize the Speaker's Goals

At this point one might ask whether this high level intentionality on the part of the system is really necessary, whether some more basic and simpler literal command interpreter might not be sufficient. There is in fact considerable evidence that when people attempt to solve problems with the aid of database management systems, they persist in expressing themselves to the machine as if it really understood what they were trying to achieve (in spite of overwhelming evidence to the contrary). Cohen, Perrault, and Allen (1981) show that:

1. People, when asking questions of a database system, expect their intentions to be recognized and responded to.
2. People expect assistance from the question answerer in order to interpret the answers, correct their misconceptions of the underlying database and choose an alternate means to fulfilling their goals when a "dead end" is reached.
3. People do more than ask questions; they give commands, comments, clarifications and descriptions of their desires. Often these are expressed in terms of a previous system response.

When examined closely, the evidence indicates that such capabilities are essential for fluent communication. In fact, there are situations in which a user cannot even express his request without first coming to some understanding with the machine about the nature of the situation. Consider the following examples from actual user interactions with the PLANES system (Waltz & Goodman, 1977), a natural language question answering system that deals with Naval aircraft maintenance:

> What was the average downtime of aircraft at each actorg? That includes maintenance time and AWM.

The user (who in this case was interfacing to the PLANES database through a person rather than a language understanding program) asked a simple question, but followed it with a comment which informed the system of what "counted" as downtime in interpreting the question. Interestingly, the PLANES database system could not take into account the user's purpose, so it printed a list of average downtimes without indicating which actorg each time was associated with.

Another type of question users ask is evidenced by the example below. The question assumes that there were F14s on the Enterprise in March. The system, however, could answer "Zero" to this question even if no F14s were on the ship — the answer would be truthful but misleading, as it fails to take into account the user's intentions.

Of the F14 aircraft based on the Enterprise in March, how many were NOR while on the ship?

A third example illustrates how users and NL systems (actually persons simulating the part of a NL system) interact, making use of the discourse context, the intentions of each other, and the ability to modify a previous request.

U: How many cases of FOD were experienced?

S: (THAT QUERY WOULD TAKE ABOUT 20 MINUTES TO COMPUTE) (IT CAN BE USEFULLY LIMITED BY CONSIDERING SPECIFIC AIRCRAFT)

U: Consider aircraft #27.

S: NIL

In this example, the system is aware of limits on the user's time and offers suggestions accordingly. The request to "Consider aircraft #27" is a vague request out of context, but here the user intends for the system to answer the question about FODs relative to the one aircraft.

A final example from the protocols of Genesereth (1978) illustrates the implicit demands that people state in conversations. The user in this example is a MACSYMA user who is having difficulty solving equation D6 and has called on an advisor for help.

User: I was trying to solve D6 for Y, and I got 0.

Advisor: Did you expect COEFF to return the coefficient of D6?

User: Yes, doesn't it?

His remark to the advisor, which on the surface just explains his difficulty, also converys his expectation that the advisor will explain why he's in trouble. The advisor's response question is understood to be part of his debugging assistance even though neither he nor the user state this explicitly.

The above examples illustrate the need for a system that can take into account the discourse context and the user's intentions in interpreting what the user wants the system to do. Constructing a system with these capabilities will require significant research in a number of areas, many of which have not been adequately studied. One of these is the need for situation dependent interpretation of linguistic devices such as "deixis" and "anaphora." The mechanism of *anaphora* permits one to make a subsequent reference to something that has previously been said in a dialog (e.g., using pronouns or definite noun phrases to refer to previously mentioned objects). *Deixis* involves similar references to things that have not been said but are present in some way in the nonlinguistic context of the conversation (e.g., in this case, what has just happened on the display screen). Anaphora has been extensively studied in linguistics (although the problems are far from solved), whereas deixis of the kind that occurs in the display context is considerably less well understood.

The resolution of both deictic and anaphoric reference requires a system to perform certain kinds of common sense inferences about the possible meanings of alternative possible referents, and to assess the plausibility of those alternatives. This in turn requires an ability to store and use considerable amounts of knowledge about the domain of discourse and the goals and objectives of the user. In addition to these linguistic devices, there is another level of interpretation of the user's input that depends even more critically on the use of such knowledge. This is the filling in of details that the commander can be assumed to have intended but did not literally say.

8. The Need for Knowledge Representation Research

The above discussion illustrates the extent to which the representation and use of (a) general world knowledge, (b) knowledge of the domain, and (c) knowledge of the goals and objectives of users are critical in the development of fluent communication and effective information display. Moreover, these problems are fundamental bottlenecks in a variety of other artificial intelligence applications. Consequently, a major portion of the BBN project has been devoted to fundamental problems of knowledge representation and use.

The KL–ONE knowledge representation system, developed during this project, has an exceptionally good representation for the inheritance relations among structured concepts, including the relationships between corresponding parts of their structures. It has been used for representing a variety of different kinds of information in our current system, and has proven to be well structured in many respects. Some of the major features of KL–ONE are:

- Inheritance of structured descriptions, i.e., when one concept is subordinate to another, the first "inherits" the properties of the second.

- Taxonomic classification of generic knowledge, i.e., when a new generic concept is introduced into the network of existing concepts, it is automatically assimilated into the network at the right place so that it inherits the appropriate characteristics of related concepts and its properties are inherited by appropriate subordinates.
- Intensional structures for functional roles, i.e., the use of a distinct class of nodes to formalize the notion of the role being played by an individual as distinct from the individual itself. This distinction allows, for example, the specification of properties of a role when the filler of that role is unknown, and allows the correct representation of situations in which a particular individual occupies several roles in an organization.
- Procedural attachment, i.e., associated with any concept or role one may record computational operations that are to be performed on instances of that concept. For example, a procedure for displaying an organization chart may be associated with the organization role of the concept of an organizational unit. Thus, when a user requests an organizational chart of a particular company, the procedure for producing the display can be inherited from the generic concept.

Knowledge representation research is one of the key problem areas in the development of intelligent natural language communication systems. There are many subtle problems of representation that are currently undergoing active investigation and whose solutions are required for really fluent natural language communication with machines. Some of the major ones have to do with the representation and reasoning about the goals and beliefs of other agents, modeling the structure of continuous discourse, modeling time and space, understanding what is required to represent and use abstract concepts, and representing the subtleties of individual identity for entities that nevertheless change their properties over time.

9. Status and Prospects

The obvious question that one might ask about a natural language understanding capability of the kind I have described is "When can we have it?" If such capabilities do not now exist, when will they? These questions, unfortunately, do not have a crisp answer. On the one hand, limited natural language understanding systems do exist today. One of the earliest successful examples of such a system is the LUNAR system, which answers factual questions about the chemical analyses of the Apollo moon rocks (Woods, 1972). Nevertheless, there are aspects of natural language understanding that have not only not yet been solved, but whose solution has

not yet been articulately envisioned. Thus, there is no point in the fore-
seeable future when the natural language problem will be totally solved.

For some applications, the existing level of capability demonstrated
by the LUNAR system is adequate — this includes the retrieval or computation of answers to specific factual questions of well-understood types
from a database of well-formatted information. It also includes the collection into a summary report of scattered pieces of data that satisfy a
common predicate. However, there are many other capabilities that are
required for a useful approximation to the kinds of capabilities that one
would like. These include the resolution of pronominal reference (and other
forms of anaphora), the resolution of ambiguity and/or vagueness in the
user's requests, and the generation of helpful responses that take into account the perceived goals of the user. These capabilities are quite limited
in current natural language understanding systems, and these are among
the areas of primary research interest today.

Capabilities similar to LUNAR's are available today in systems such
as Larry Harris's Intellect system and others described in this conference.
Moreover, there is likely to be a continued gradual incorporation of more
discourse oriented capabilities into such systems. A major qualitative difference will occur, however, when systems become available that have a
good conceptual model of the application domain with the same conceptual
structure as that of the human user (as opposed to the data structure conceptualization typical of most data base management systems). The advent
of such systems, however, will require further research in knowledge representation, knowledge-based inference, common sense reasoning, and
recognition of user's plans.

10. *Conclusion*

The need for natural language understanding systems arises in many different contexts, including computer-based decision making. Particularly in
crisis situations, a decision maker must be able to interact with an information processing system as easily as with humans, and in a language that
matches his own cognitive abilities. The system must be able to handle
incomplete or inexact requests for information, resolve ambiguities, and
use its knowledge of the domain of discourse and current context to understand pronomial and other indirect references. Although it might be
possible to design an artificial language that would meet the above needs,
the preferred method for developing such a capability is to use natural
English.

The primary advantages of English for such applications are:

1. The user need not learn and remember a large number of special
   conventions for communicating with the system.

2. The underlying conceptual structures of English can match the user's conceptualization of the task being performed.
3. One can use English to express instructions at varying levels of detail.
4. English includes shortcut devices such as anaphora (using "it", "the", etc. to refer to objects or phrases previously mentioned) and deixis (using nonlinguistic devices, such as pointing to an object on a display screen) to specify one's intent.
5. The conceptual structure of English allows the incorporation of abbreviations and other shorthand devices for concisely expressing frequent commands.

Systems that allow natural language access to a conventional computerized data base are becoming available as software products on today's markets. Such systems generally have limitations in the range of English syntax that they will understand and severe limitations in their discourse understanding abilities. Moreover, they are dependent on the generally artificial conceptualizations of the domains that are built into their data bases. In the next few years, such systems can be expected to evolve somewhat more sophisticated discourse understanding abilities (better focus of attention models and handling of anaphoric reference), but will still fall short of intelligently understanding what the user wants and responding appropriately. Systems with the latter capabilities will emerge as further progress is made in knowledge representation, modeling belief systems, and common sense reasoning.

11. *References*

Brachman, R. J., Bobrow, R. J., Cohen, P. R., Klovstad, J. W., Webber, B. L., & Woods, W. A. *Research in natural language understanding — Annual report: 1 Sept 78 – 31 Aug 79* (BBN Report No. 4274). Cambridge, MA: Bolt Beranek and Newman Inc., August 1979.

Cohen, P. R., Perrault, C. R., & Allen, J. Beyond question answering. In W. Lehnart & M. Ringle (Eds.), *Strategies for natural language processing.* Hillsdale, NJ: Lawrence Erlbaum Associates, 1981.

Genesereth, M. R. *Automated consultation for complex computer systems.* Unpublished doctoral dissertation, Department of Computer Science, Division of Applied Sciences, Harvard University, 1978.

Sidner, C. L. *Protocols of users manipulating visually presented information with natural language* (Tech. Rep. 5128). Cambridge, MA: Bolt Beranek and Newman Inc., September 1982.

Waltz, D. L., & Goodman, B. A. Writing a natural language database system. In *Proceedings of the Fifth International Joint Conference on Artificial Intelligence, Vol. 1.* 1977, 144–150.

Woods, W. A., Kaplan, R. M., & Nash-Webber, B. L. *The lunar sciences natural language information system: Final report* (BBN Report 2378). Cambridge, MA: Bolt Beranek and Newman Inc., June 1972.

In Response: Next Steps in Natural Language Interaction

Bonnie Lynn Webber and Tim Finin
DEPARTMENT OF COMPUTER
AND INFORMATION SCIENCE
UNIVERSITY OF PENNSYLVANIA
PHILADELPHIA PA 19104

In the area of man-machine interaction, Natural Language has so far primarily been used to simplify people's access to information. The next step beyond simple data access is the kind of cooperative interactive problem-solving that current expect systems aspire to. But support for problem-solving (which includes helping the user formulate his/her problems) demands more in the way of interaction than just answering requests for factual information. In the first part of this paper, we illustrate some of these needed capabilities. In the remainder, we discuss two of them in greater detail: (a) recognizing and responding to user misconceptions and (b) getting from users the information needed to help them solve their problems.

1. Introduction

In the area of man-machine interaction, Natural Language has primarily been seen as a significant way of simplifying people's access to system services and information. Potential users need not spend time learning or trying to remember some arcane formalism: they can express their requests as they would everyday.

This viewpoint has led to valuable research on removing what appear to be arbitrary, artificial constraints on a user's freedom of expression: parsers can handle most of English syntax (Bobrow, 1978; Robinson, 1982); domain-specific processors can be tuned to interpret most reasonable utterances within the domain (Shwartz, 1984); utterances can be interpreted to some extent in the context of the previous discourse (Grosz, 1977; Hendrix, et al., 1978; Sidner, 1982; Waltz, 1978; Webber, 1978) and in light of the user's underlying intentions (Allen, 1982; Cohen, et al., 1981; Sidner,

1981); misspelling and grammatical errors can be tolerated to some extent (Hendrix, et al., 1978; Kwasny & Sondheimer, 1981); and in combined graphics/Natural Language systems, utterances can be combined with pointing for added naturalness (cf. Woods, 1984).

On the other hand, this viewpoint seems to assume that users know what information they want and can use it to solve their problems themselves. The former, unfortunately, is not always the case, and the latter is not always possible. In fact, the frequent lack of in-house expertise needed for solving problems is often cited as the reason for developing "expert systems" in the first place.

But support for problem solving (which includes helping the user **formulate** his/her problems) demands more in the way of interaction than just answering requests for factual information. In the first part of this paper, we illustrate some of these needed capabilities. In the remainder, we discuss two of them in greater detail to show the kind of system support they require.

Our point is that if we are to go beyond simple data access via Natural Language to the next step — cooperative problem-solving interactions — we must look at the system's role in the interaction. If we fail to recognize this complementary issue, much of the advantage of Natural Language input will be lost.

2. Problem-Solving Interactions

What we aspire to is the type of discourse behavior displayed in cooperative interactive problem solving among humans. To characterize such interactions, two of our students (Pollack, et al., 1982) have collected and analysed transcripts of a "naturally occurring" expert system — the radio talk show "Harry Gross: Speaking of Your Money" (WCAU, Philadelphia). In this program, listeners call in to ask for financial advice, which the expert, Harry Gross, attempts to provide. The ensuing discourse is basically a cooperative problem solving interaction. While not a perfect model for machine-based expert systems, the talk show transcripts do suggest many types of interactions that these systems should endeavor to support.

An examination of these transcripts reveals a regular pattern of interaction, rather like a negotiation — the process through which people arrive at a conclusion by means of a discussion. Rarely does a caller simply state a problem and passively listen to the expert's response. Rather what ensues is a collaborative dialogue in which caller and expert negotiate to determine a statement of the problem the caller *wants* solved and the expert *can* solve, and the statement of a solution the expert can support and the caller can accept and ideally understand.

More specifically, we have noticed such activities as the following during these negotiations. ('H' stands for the expert, Harry Gross, and 'U', for the caller.)

- The user attempts to verify his/her understanding of what the expert has said and the expert responds to either confirm or clarify — e.g.,

    H: Okay, in your case I would not object to see you go back into that for another 6 months.
    U: So you roll it over, in other words?
    H: Right.

- The user suggests an alternative solution to that proposed by the expert, and the expert responds either to confirm its possibility or to show why not — e.g.,

    H: Put the money aside in T-notes.
    U: Now wait. In a 43% bracket I didn't think that would be wise. I thought maybe we should buy municipal bonds.
    H: If you buy municipals, the interest on your loan won't be deductible. So municipals just don't make sense.

- The user requests justification of the expert's suggestion, and the expert provides it — e.g.,

    H: You can stop right there. Take the money.
    U: Take the money?
    H: Right. You're only getting $1500 a year. At $17,000, no trouble at all to get 10% on $17,000.

- The caller shows confusion about a term or concept or explicitly requests information on it, and the expert provides clarification — e.g.,

    H: I'd like to see you put that into two different Southern utilities.
    U: Southern utilities?
    H: Yes.
    U: Huh?
    H: Utilities that operate in the South — Texas, Oklahoma, Florida, Georgia, . . .

- The expert requests information from the user and then works with him/her to get it — e.g.,

> H: Have you gains on other securities?
> U: Yeh, we have some certificates and real estate.
> H: Sorry, I'm not making myself clear — if you had some other
>   stock
> U: Oh
> H: Do you have any paper gains? On other securities?

- The user asks or answers a question, states a preference, etc. that
  shows a misconception, and the expert attempts to correct it, instead
  of or in addition to responding to his/her utterance. (Examples of
  this activity will be given in section 3.)

While these and other activities are discussed in more detail in Pollack, et
al., (1982), their relevance here is that if the system is not capable of reacting
appropriately in such interactions, the user may become confused by what
the system **does** do in response. For example, consider utterances of the
form "What about <x>?". At least two major systems (Hendrix, et al., 1978;
Waltz, 1978) treat all such utterances as a short way of asking a parallel
question to one asked earlier — e.g.,

> U: What is the length of each Russian aircraft carrier?
> S: 420 feet
> U: What about the draft?

The second question is correctly taken to mean "What is the draft of each
Russian aircraft carrier?" But utterances of that form can in fact also be
used for another purpose mentioned above — to propose an alternative
possible answer to the one given by the system. For example,

> U: What's a good thing to invest my pension in?
> S: Put the money aside in T-notes.
> U: What about municipal bonds?

It would be devastating to treat the user's second question as a request
for a good thing in which to invest municipal bonds! The conclusion is that
systems must be able to perform additional functions (such as considering
the user's proposed alternative) and recognize when they are called for.

A growing number of researchers are involved in developing the capabilities needed for Natural Language problem-solving interactions. This
work includes that of Swartout (1981) and Wallis and Shortliffe (1982) *explanation,* McKeown (1982) *term definition,* and Wahlster and his colleagues
(van Hahn, et al., 1980), Wilensky (1982), Woods (1984), Schank and Slade
(1984), and Shwartz (1984) *advisory system structure.* The interactional capabilities we will be discussing in the remainder of the paper are not covered

in this other work: section 3 focusses on recognizing and responding to user misconceptions, and section 4, on getting from the user the information a system needs to help the user solve his/her problems.

What we hope to gain by this discussion, besides the reader's increased awareness of the importance of the system's role in the interaction, is recognition of the fact that in order to perform at this more sophisticated level, systems need **both** enrichments to existing data models or knowledge representations **and** additional types of reasoning. In other words, appropriate interactive behavior will not come about merely by tacking onto an existing system some off-the-shelf front end. The mechanisms that bring it about will have to tie deeply into what must already be rich and powerful representation and reasoning components.

We want to emphasize also that spontaneous automatic generation of fluent Natural Language, though a desirable goal, is not the primary point here. The interactional capabilities that we discuss are needed even if a system is accessed using a formal notation: database systems will have to be as careful not to mislead users by their responses (Webber, 1983), while even with canned text, expert systems will have to make similar provision for getting their questions answered.

What is important — and this we get by looking at Natural Language interactions as our model — is that an interactive system, whatever it is, must follow everyday conversational principles and practices. If it does, a user's **normal expectations** about responses to his/her utterances and **normal strategies** for interpreting those responses will not fail him/her, even though the conversational partner is a machine.

3. Recognizing and Responding to Misconceptions

Much of our knowledge of the world is incomplete; a lot of it is faulty. Much of the time, it makes no difference. At times though, it does — for example, when trying to solve a problem or acquire some information. At those times, misconceptions may lead one to try to solve the wrong problem, to seek an inappropriate solution or to misunderstand and hence be misled by the information one receives. It is the latter point that the work described here addresses. User misconceptions about the domain or its encoding in the database or expert system may lead the user to draw false conclusions from the system's response to his/her question. The system must do what it can to prevent this. We shall discuss four types of user misconceptions here: (a) misconceptions that something exists (or that the system knows of its existence), (b) misconceptions that something can participate in some relation, (c) misconceptions about the classification or properties of some object, and (d) misconceptions that some event can occur.

3.1. *"Extensional" Misconceptions*

As many people have noted, most database queries can be considered requests for the extension of some set descriptor (i.e., a listing of the individuals satisfying that description). Such descriptors are made up by restricting descriptions of larger sets in various ways. For example, the question

> Which foreign-born employees work in the shoe department?

can be taken as a request for the set of individuals satisfying the description "foreign-born employee who works in the shoe department". This is composed of the more inclusive descriptor "employee" restricted to those who are foreign-born, restricted again to those who work in the shoe department. One obvious misconception that a user can hold in asking such a question is that some description has a nonempty extension in the database, when in fact it doesn't. In that case, the answer to the user's question will follow trivially, without the user realizing it. If the system can instead recognize and point out such misconceptions, the user will be better off. This is the aim of the CO–OP system developed at the University of Pennsylvania in 1979 (Kaplan, 1982).

For example, a question like "Which French majors failed CIS531 last term?" reveals *inter alia* the questioner's belief that there are French majors. If there are in fact no French majors, an unqualified "None", while technically correct, would confirm the questioner's false belief. What are the consequences of unintentionally confirming such beliefs if they are incorrect? If the questioner concludes from "None" that no French majors failed CIS531 last term, s/he might in turn believe that all French majors **passed** CIS531 last term, and in turn many more unwarranted things about the abilities of French majors.

CO–OP detects such misconceptions in the course of retrieving answers to the database query viewed as a composite set descriptor. If one of the more inclusive subset descriptions making up the query is found to have an empty extension, then search halts and the user is informed about the system's lack of knowledge of individuals satisfying the failing descriptions — e.g., "According to this database, there are no French majors." The user can then take this information into account in formulating further queries.

3.2. *"Type" Misconceptions*

A second common type of misconception is that some entity or subset of entities can participate in some relation. This is similar to type violations in programming languages, where, for example, a function or procedure call may be incorrect because its arguments are of the wrong type. Some initial work in this area is reported in Mays (1980). The knowledge needed to recognize type failures in users' queries is contained in the system's database schema and consists of entity-relation information, hierarchical (subset-superset) information, as well as partition information as to what collection of subsets of a given set are mutually exclusive. It is the last factor that is critical for distinguishing between a nondeviant request like

>Which women teach courses?

and a deviant one like

>Which undergraduates teach courses?

where — as shown in Figure 1 — the TEACH relation is asserted to hold between FACULTY and COURSE. As Figure 1 also shows, the entity PEOPLE is partitioned in two different ways — into MEN and WOMEN, and into FACULTY and STUDENT. Thus if an entity is classified as a MAN, it cannot also be classified as a WOMAN. But it can also be classified as FACULTY or STUDENT (but not both). Assuming a relation is always asserted at the most general point in the hierarchy, the meaning of the configuration is taken to be that **only** FACULTY teach COURSEs. STUDENTs cannot TEACH COURSEs, nor can any subset of STUDENT do so. Hence the TEACH relationship cannot hold between UNDERGRADUATE and COURSE. The same

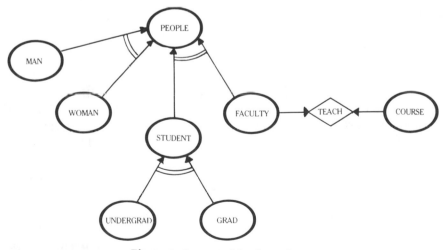

Figure 1. *Segment of database schema.*

is **not** true of WOMEN, as entities can be classified as both WOMEN and FACULTY.

In the case of type failures that it recognizes, Mays' system does not answer the user's query directly but rather provides relevant information about the entities and relation involved. That is, if the misconception involves a relation R holding between entities A and B, the system seeks relevant information by abstracting on each of R, A and B and determining whether either RA_ or R_ B or _AB can hold in the database for some value of "_". For example, the system's response to the above question "Which undergraduates teach courses?" would be

> S: I don't believe that undergraduates can teach courses.
>   Faculty **teach courses.**
>   **Undergraduates** take **courses.**

Further work is needed on several problems involved in recognizing and responding to "type" misconceptions: first, in the context of negative questions — e.g.,

> U: Which courses are not taught by faculty?
> S: I believe that only faculty can teach courses. Therefore all courses
>   are taught by faculty.

The user's question reveals an incorrent belief that the TEACH relation can hold of some entity *other than* FACULTY. May's system would not catch this misconception because it only tries to confirm the explicitly mentioned types — i.e., FACULTY, COURSE and TEACH. Further work is also needed on misconceptions involving "complex" types (requiring a more highly structured knowledge representation than Mays' system employed, for example Kl–One (Schmolze & Brachman, 1982)) — e.g.,

> U: How many widgets did we buy from Ajax last year?
> S: Ajax doesn't supply widgets — they supply clocks.

The relation BUY holds of a CUSTOMER, a SUPPLIER and a PRODUCT, and all these simple type constraints are satisfied in the question. However, there is in fact an additional constraint that the PRODUCT BOUGHT from a SUPPLIER is part of the PRODUCT LINE of that SUPPLIER. One would like to enable a system to apply this constraint both in standard updating procedures and as a way of recognizing and responding to "type" misconceptions.

3.3. *"Object-related" Misconceptions*

A third type of user misconception involves concepts modelled by the system. That is, there is some discrepancy between what the user believes about an object and what the system believes about it. This discrepancy may be revealed in a question the user asks the system — e.g.,

> U: What is the interest rate on this stock?

or in his/her response to a question asked by the system — e.g.,

> S: Do you have any liquid assets?
> U: I have a $5K money market certificate.

As above, such discrepancies can either prevent the system from answering the user's query or prevent the user from correctly understanding the system's response or prevent the system from getting the information it needs (see also section 4). Thus the system should respond to the discrepancy as effectively as possible. Enabling it to do so in a domain-independent way is the aim of current research here being done by McCoy (1983).

When a person in conversation with another makes a error that may be the result of a misconception (as opposed to just a mistake), the very least the person expects from his/her partner is to have the false information denied — e.g.,

> U: What is the interest rate on this stock?
> S: Stocks don't have an interest rate.
>
> S: Do you have any liquid assets?
> U: I have a $5K money market certificate.
> S: A money market certificate isn't a liquid asset.

However, through studying transcripts of people interacting with various experts (Malhotra, 1975; Malhotra & Sheridan, 1976; Pollack, et al., 1982; Schuster, 1982), McCoy has identified additional types of information often included in such responses — for example, a true statement that stands in contrast with the incorrect one — e.g.,

> U: What is the interest rate on this stock?
> S: Stocks don't have an interest rate. Stocks may pay a dividend.

or information aimed at convincing the user of the correction or at pointing out and refuting the faulty reasoning perceived to have been done — e.g.,

> U: What is the interest rate on this stock?
> S: Stocks don't have an interest rate. Only investments which have a face value (like bonds) have an interest rate.
>
> S: Do you have any liquid assets?
> U: I have a $5K money market certificate.
> S: A money market certificate isn't a liquid asset. Your money is tied up for a period of several years in a money market certificate.

or even information that suggests a correction — e.g.,

> S: Do you have any liquid assets?
> U: I have a $5K money market certificate.
> S: A money market certificate isn't a liquid asset. Perhaps you meant
> a money market fund.

McCoy's problem is two-fold: (a) to characterize in a domain-independent fashion what influences the choice of additional information to include in a given correction response and (b) to enable a system to produce such responses automatically. The former involves identifying possible misconceptions that may lead to such an error, while the latter involves establishing and maintaining a model of things the user correctly knows so as to eliminate them from consideration. One example must suffice (for additional discussion, see McCoy, 1983):

> S: Do you have any liquid assets?
> U: I have a $5K money market certificate.
> S: A money market certificate isn't a liquid asset. Your money is tied
> up for a period of several years in a money market certificate.
> Perhaps you meant a money market fund.

If the user's response is understood as an affirmative answer to the system's question, it must be the case that s/he believes s/he has something that can or might be classified as a "liquid asset." However, a money market certificate isn't one. Thus the user's inappropriate response might follow from either:

1. his/her lack of knowledge as to what a liquid asset is (i.e., his/her response might just be a guess, naming some investment s/he knows s/he has);
2. his/her incorrect belief that a money market certificate is a liquid asset (i.e., s/he believes that it has the attributes of a liquid asset);
3. his/her incorrect belief that what she has is a money market certificate (i.e., s/he is confusing what s/he has with a money market certificate).

If the system has no reason to rule out any of these possibilities, it might try to respond to all of them. Thus it addresses point 2 explicitly in the first sentence of its reply and point 1 implicitly in the second (i.e., by contrast, a liquid asset is one that is not tied up for a long period). In its third and final sentence, the system addresses point 3, thereby hitting all bases. Remember that it is still the system's goal to determine if the user has any liquid assets. If his/her money is in a money market fund, then the answer is "yes". This is the type of behavior that McCoy is attempting to support.

3.4. "Event-related" Misconceptions

The fourth type of misconception involves events and states and their dependencies over time. It is possible for a user to be mistaken about what
can be true now or what could have been true (or happened) in the past,
because (a) s/he is unaware of the occurrence of some event or (b) s/he
does not know that an event has particular consequences or (c) s/he believes some event has occurred when it hasn't. Again, given a question
revealing such a misconception, a simple "no" or "none" answer would
just perpetuate it. For example,

> U: Is John registered for CSE220?
> S1: No.
> S2: No. He can't be registered for it because he has already advance
>   placed it.

For the user to have asked this question, s/he must be ignorant of either
the event that precluded registering or its consequences. While the system
may not be able to determine which, it should provide the user with enough
information that s/he can square away the misconception him/herself. (In
response S2 above, the general rule "Advance placing a course precludes
registering for it", which the user may in fact be ignorant of, is not stated
explicitly but is assumed to be derivable from the response.)

The knowledge needed to recognize and square away such misconceptions consists of a knowledge of past events (or states of the database)
— often preserved but not accessible to the database system — and of the
relationship between past events and what can be true afterwards, including
possibly the present. The latter is very much like update constraints used
to maintain detabase consistency. However in general, update constraints
are not expressed in a form that admits reasoning about possible change.
Something more is needed. What we have chosen to use in our own research
is an extension of the propositional branching time temporal logic (BenAri,
et al., 1981), as documented in Mays (1982, 1983).

Our original impetus into this area was a desire to give a database
system the ability to take the initiative and offer to monitor for information
of which it was currently unaware. For example,

> U: Has John checked in yet?
> S: No — shall I let you know when he has?
>
> U: Has John checked in yet?
> S: Yes — shall I let you know when the rest of the committee members
>   do?

Work on producing monitor offers that are both competent (i.e., that correspond to a possible future state of the database) and relevant (i.e., that

the user would be interested in) is proceeding concurrently with the work reported on here (Mays, 1982, 1983). We have termed systems which can reason about possible future states of the database "dynamic database systems".

We do not have the space here to explain in detail the logical system we are using (but see Mays, 1982). Briefly, there is a reserved time point NOW, where the past is viewed as a linear sequence of time points (prior to and including NOW) and the future is viewed as a branching structure of time points (following and including NOW). A set of nine complex operators are available to quantify propositions as to the points they are asserted to hold over — e.g.,

> ∀Gq—proposition q holds at every point of every future
> ∃ Gq—proposition q holds at every point of some future
> ∃ Xq—proposition q holds at the next point in some future
> Pq—proposition q holds at some time in the past
> etc.

Axioms assert the relationship between events/states in the past, present and future. For example, if the propositional letter 'a' is taken to stand for 'student advance places course' and 'r', for 'student is registered for course', the following axiom states the continuing rule that a student who has advance placed a course (some time in the past) is not now registered for it —

$$\forall\ G(Pa \rightarrow \sim r)$$

For the above registration example, to distinguish whether it is **accidental** that John is not registered for CSE220 now (he could be, only he's not) or **foreordained** (some event has taken place that precludes registering or some enabling event has not yet occurred) requires the system to suppress its knowledge (or assumption) about John's current status[1] and consider whether it could provably believe the opposite — i.e., that John is registered now for CSE220. If it couldn't, then not only is John **not** registered for CSE220, the system should have identified at least one basis for why he couldn't be. By the above axiom, it is clear that it is not accidental that John isn't registered, because r (being registered for CSE220) is inconsistent with Pa (having advance placed CSE220), that is, Pa → ~r.

In summary, we consider it very important for the development of the

---

[1] which may involve suppressing other facts that follow from his being registered — e.g., that he is taking four courses, that he is in Room 225 from 10–11am MWF, etc. A data dependency scheme such as that discussed in Doyle (1979) or McDermott & Brooks (1982) would be needed to do so.

next generation of cooperative interactive problem-solving systems to enable them to recognize and respond to user misconceptions. In addition to the work discussed above, we have started lines of inquiry into detecting and correcting other types of misconceptions — in particular,

- user misconceptions about how to carry out some procedure (Schuster, 1983), for use in smart Help Systems — for example,

  > U: How do I invoke the editor?
  > S: You don't have to invoke it. Whenever you're not running some other program you're in the editor.

- user misconceptions about the "best" course of action to achieve some goal (Pollack, et al., 1983).

  > U: Do you think it's better to buy T-bonds now or wait a month or two?
  > S: In your bracket, you'd be better off buying municipals.

Clearly not all misconceptions can be detected, not all can be easily corrected and, moreover, not all make a difference to the system's ability to convey its information responsibly (Joshi, 1982). What we believe AI can contribute to this area is an improvement on a system's ability to detect and correct misconceptions detrimental to the successful transfer of information from system to user.

4. Getting Information from Users

As we noted in the Introduction (section 1), systems must be able to get from their users the information they need to help these users solve their problems. The most commonly used way of getting information is via "menus" — essentially, multiple choice questions. However, there are several problems with relying on menus:

- The user may not understand either the question or the menu options.
- The user may be influenced by the options — i.e., s/he assumes one of them must be appropriate to his/her case, so s/he bends the facts to fit the options.
- The user may not be satisfied with any of the options — i.e., none seems appropriate to his/her case.
- The user may want to qualify his/her response — i.e., s/he may feel that simply agreeing to a particular option will be misconstrued.

In all these cases, the reliability of the user's response is called into question.

For these reasons, we are attempting to provide users with as much freedom in **responding** to questions, ideally in Natural Language, as other systems provide users in **asking** them. This involves at least the following:

1. allowing users more leeway in how they provide the requested information, along with any additional information they believe relevant and want to convey as well.
2. providing the user with more help when s/he doesn't know how to respond.

The first point is important because, ordinarily, people vary in how they respond to questions. Some are direct, some are roundabout, some say a lot, some say a little, some do not know what to say. If systems cannot accommodate this diversity, they will drive users away. On the other hand, there are general patterns within this diversity. For example, one general response strategy involves not only answering the given question but, along with it, providing additional information believed to be relevant to the current (shared) task — e.g.,

> S: Would you like to sit in smoking or nonsmoking?
> U: Non-smoking, on an aisle, near the front please.

As van Katwijk, et al., (1979) has shown experimentally, people are very annoyed if this additional information they offer is ignored. While it is already the case that several frame-based systems (including Bobrow, et al., 1977; Engleman et al., 1980; and Shwartz, 1984) accept volunteered information, there are other significant question-response patterns that a system should also be prepared for.

In the first part of this section, we discuss various common ways in which people respond to questions and the relationship between (a) their **response,** (b) the **answer** to the question and (c) the **additional information,** if any, being offered. We also describe some research being done here whose aim is to enable systems to accommodate one of these response strategies that we see as particularly important. In the second part of this section, we discuss various ways of helping a user who doesn't know how to respond to a question.

4.1. *Responses from which an Answer can be Inferred*

Often a person will respond indirectly to a question, in the belief that the questioner can infer the answer from the response. (We believe the same will be true of users responding to system questions.) We have identified four situations in which this occurs.

1. The user is **unable** to determine an answer to the question but has what s/he believes to be information from which the system can deduce an answer — e.g.,

> S: What is your employee classification: A–1, A–2 or A 3?
> U: I'm an assistant professor in Oriental Studies.
> S: All faculty members are A–1 employees, thank you.
>
> S: Are you a senior citizen?
> U: I'm 62 years old.

Of course the user can be wrong, and the information s/he offers either inadequate or irrelevant — e.g.,

> S: What is your employee classification: A–1, A–2, A–3?
> U: I've been here for over 5 years.
> S: Sorry, could you tell me either your job title or position?

2. The user is **able but unwilling** to perform the computations necessary for answering the question. Instead s/he provides data that s/he believes the system can use to compute an answer — e.g.,

> S: What is your yearly salary?
> U: $1840 per month

3. The user decides to be more informative than the question calls for, responding with an instantiation of a general description contained in the question — e.g.,

> S: Did you see any fish?
> U: I saw some guppies.

4. The user feels that a direct answer to the given question would be **logically correct but misleading.** That is, it would imply that there is nothing more s/he can say, relevant to the situation prompting the question. Instead s/he provides **related** information, whose relation to the question s/he believes the questioner can and will recognize — e.g.,

> S: Did you invite the Bennets?
> U: I invited Elizabeth.
>
> S: Did you invite Elizabeth?
> U: I invited all the Bennets.

Neither reply answers S's simple yes/no question directly. However, a bare "no" in the first case and a bare "yes" in the second, while literally true, are likely to mislead S as to what is actually true.[2] The given response is more cooperative and moreover more efficient than an explicit statement of both the answer and the added qualification.

The fourth question-response pattern is particularly important for a system to accommodate because we see a user's *fear of being misunderstood* by a computer as being of greater consequence for the future of interactive problem-solving than a user's *annoyance at volunteered information being ignored.* For this reason, it is a current topic of research (Hirschberg, 1983).

As we noted above, the interpretation of responses which fit this pattern requires that the questioner recognize the relation between the particular entity, event or situation referenced in the query and that referenced in the response. From this relation, the questioner can determine not only the respondent's answer to the given question but also the additional information the respondent has included. Two forms of reasoning are involved in determining these two things: standard logical deduction and an extended form of conversational implicature[3] recognized first by Horn (1978) and termed by Gazdar (1975) "scalar quantity implicature".

Consider a speaker whose utterance can be interpreted in terms of an open proposition **P** holding of some scalar value **x.** Horn observed that, following Grice's **Maxim of Quantity** — make your contribution as informative as is required (for the current purposes of the exchange) — the speaker in effect commits him/herself to **x** being the highest value on its scale that **P** holds of, if s/he is observing Grice's **Maxim of Quality** — Do not say what you believe to be false. Propositions formed from **P** by substituting for **x** values higher on the scale (which Horn limits to those that logically *entail*[4] **x**) are thereby implicitly marked — i.e., **implicated** — by the speaker either as 1) not **known** to be the case or as 2) known **not** to be the case, depending upon the discourse context.

A brief example: the lexical items "some" and "all" can be seen as

---

[2] It seems clear to us that the third and fourth response patterns are related. On the other hand, it seems that they can be distinguished by how the response can be paraphrased. Responses following the third pattern can always be paraphrased "Yes, specifically . . .", as in "Are you 65 or older? Yes. Specifically, I'm 72". Not so for the fourth pattern: its correct paraphrases include "No, but . . ." and "Yes, moreover . . .", as in "Did you invite the Bennets? No, but I invited Elizabeth" or "Did you invite Elizabeth? Yes. Moreover, I invited all the Bennets".

[3] An **implicature** is a conclusion that a person would draw from an utterance over and above its propositional content or anything that proposition logically implies. The concept of implicature comes from Grice (1975).

[4] Horn's definition of logical entailment is two-sided: **P** entails **Q** iff $p \rightarrow q$ and $\sim q \rightarrow \sim p$.

points on a "quantifier" scale, in which "all" entails "some". Thus in the sequence

> A: Did anyone leave early?
> B: Some people did.

the use of "some" not only explicitly communicates that some people left early but implicates that not all of them did. If all of them did, the cooperative speaker would have committed him/herself to the higher point on the scale. The implicature associated with an utterance of the form "Some X's Y" is thus that "Not all X's Y".

Research on this fourth question-response pattern is being done here by Hirschberg (1983). To begin with, she has extended scalar quantity implicature to include *hierarchical* ordering relations between entities, events and situations (e.g., member-set, part-whole, subtype-type) as well as linear ordering relations, not only entailment but also temporal orderings — both of events and of states — and spatial orderings. For example, in the first exchange above

> S: Did you invite the Bennets?
> U: I invited Elizabeth.

the two entities (the Bennets and Elizabeth) stand in a set-member relationship. The answer to the system's question follows by implicature: questioning whether an open proposition holds of the set and responding by specifying it holds particular members of that set implicates that the proposition only holds of those members. Next consider

> S: Have you dealt the cards yet?
> U: I shuffled them.

Here the two events (dealing cards and shuffling them) stand in a linear temporal relationship (i.e., within the process of playing cards). Again the answer to the system's question follows by implicature: questioning one stage in the process and responding with a prior stage implicates that the process has only been taken as far as that prior stage. The latter stage has not yet been reached.

In the examples so far, the answer to the system's question follows by *implicature* from the user's response. Hirschberg has noted other reasoning strategies involved as well. For example, in contrast to the first example above, the question may be whether some property holds of a set member and the response may assert that it holds of the entire set — e.g.,

> S: Did you invite Elizabeth?
> U: I invited all the Bennets.

Here, the answer to the given question follows from the response by *deduction,* not by implicature (in particular, by "universal" instantiation: if ∀x.Px then Pa). The additional information is that the other members of the Bennet family besides Elizabeth have been invited as well.

This reasoning can get quite complex, employing both deduction and implicature. For example, compare the following two superficially similar exchanges

S1: Did you clean your room?
U1: I made the bed.

S2: Did you clean your room?
U2: I washed the dishes.

The answer in the first exchange follows from the simple whole-part implicature mentioned above. (That is, bed making is part of bedroom cleaning. Hence the answer to S1 is "No, except for the bed".) However, dish washing is clearly NOT part of bedroom cleaning: the answer to S2 is NOT "No, except for the dishes". On the other hand, dish washing is part of house cleaning in general. If the respondent takes this as the question to respond to, his/her response can be understood by implicature as dish washing being the only part of house cleaning that the respondent did. S/he did not do anything else. Hence it follows by deduction that s/he has not cleaned his/her room.

Hirschberg's current research is focussed on developing techniques both for **recognizing** instances of this question-response pattern, using the system's domain knowledge and a model of the user's and system's mutual knowledge, and for **deriving** both the requested information and any additional relevant information from the given response.

This task is made difficult by three things. First, as the above room-cleaning/dish-washing example shows, the process may involve figuring out what question it is that the respondent has chosen to respond to.

Another difficulty is that the second argument to the comparison may not be explicit in the response, but rather may have to be **deduced** from it — as in

S: Did you see the play?
U: I didn't arrive until the intermission.

This can be understood in terms of a whole-part relationship between seeing the play and seeing the first act. (That is, by denying the proposition "seeing X" holds of one part, it follows by implicature that it does hold of the others.) However, that the respondent has not seen the act prior to intermission must be inferred from his/her response. That is,

> s/he did not arrive until intermission $= => $
> s/he was not present before intermission $= =>$
>   s/he could not see the act(s) prior to intermission (since being present is a precondition for seeing it)

The third problem is, computationally, the amount of mutual knowledge this type of question-response pattern can draw upon. For example,

> S: Did you run to Broad Street?
> U: I got to the river.

Here, the direct answer "no" is conveyed by an implicature that depends upon the participants' mutual knowledge of a linear temporal relationship between running to Broad Street and running to the river — specifically, the knowledge that West Philadelphia runners often make a loop that goes east to some point before coming back. Broad Street is east of the Schuykill River. Thus getting to the river implicates not getting further east—i.e., not getting to Broad. While this is a straightforward instance of an answer following from implicature based on a linear temporal ordering of events, recognizing it as such does demand a great deal of domain-specific knowledge that can be assumed to be mutually known.

Despite these problems, the importance we see in enabling systems to understand this type of response to their questions makes it imperative that some sort of solution (preferably a good one!) is devised.

4.2. When the User Cannot Answer

When a system makes an attempt to get information from its user, it is very possible that the user will not be able to answer. (As expert systems begin to be consulted by a broader population, their designers will be able to make fewer and fewer assumptions about the knowledge and capabilities of any individual user. Thus it becomes more and more likely that not every user will be able to answer every question s/he is asked.) A flexible system should be able to respond with appropriate information to help the user out of his/her predicament (preferably in the form of a restatement of the question so that the user knows how to respond). There are many reasons why a user might not be able to respond as intended to a question posed by the system:

- The user **doesn't understand** the question — e.g., it contains unfamiliar terms or concepts. Here the question should be rephrased in a way that conveys the meaning of the unfamiliar terms or concepts.

- The user **isn't sure** that the question means the same to the system as it does to him/her. (That is, it contains familiar terms that the user suspects may be used in an unfamiliar way.) In this case, the user needs some alternative description of the question that clarifies it.
- The user understands the question but **doesn't know how** to go about determining the answer. In this case, the system should be able to suggest one or more procedures for doing so.
- The user understands the question but **doesn't remember** the information needed for an answer. In this case, the system should be able to ask the user for information that provides strongly suggestive clues to the information it needs.
- The user understands the question but **doesn't have at hand** the information needed for determining an answer (e.g., lab results). The system should be able to figure out whether it might be able to perform some preliminary reasoning without the missing information and finish things off when it is provided.
- The user **doesn't know why** the system wants the information and won't divulge it until s/he does. In this case, the question needs to be expanded to include the system's reason for asking it.
- The user **doesn't believe** the requested information is **relevant** to solving the problem and is trying to force the expert to adopt another line of reasoning. In this case, the system should be able to identify and pursue an alternative strategy, if one exists. If not, it should be able to explain to the user why it cannot continue. The same holds if it is the case that the user does not wish to divulge the information.

Each of these situations poses a slightly different problem for a co-operative interactive problem-solving system. As things currently stand, rule-based systems built in the Mycin framework (Davis, 1979) allow a user to request justification for why the system has asked its question (WHY?). The system responds with a description of the rule whose left-hand side it was trying to instantiate at the time of asking the question. In addition, work done by McKeown (1982) can enable a system to respond to user requests for term definitions and could be extended to enable the system to rephrase its question in a way that conveys the meaning of the unfamiliar terms. What follows then is a proposal for the set of options a user should have at his/her disposal when s/he is unable or unwilling to answer the system's question, as well as the system knowledge and capabilities required to provide such options:

1. alternate descriptions of terms — e.g.,

> S: Are there other hereditary illnesses that I haven't asked about?
> U: WHAT_IS? "hereditary illness"
> S: Are there any illnesses that pass from parent to child in your family?

2. alternate conceptual models of what those terms might mean to people coming from different backgrounds or levels of expertise — e.g.,

> S: Have you been coughing long?
> U: CLARIFY "coughing long"
> S: Have you been coughing for over three days?

3. procedures for the user's acquiring the information, annotated by the level of resources, time and sophistication needed to carry them out. (This can also be used for guessing which procedures, if any, the user may have already tried, so as to avoid suggesting them.) — e.g.,

> S: Does eating certain foods cause a problem for you?
> U: HOW_KNOW?
> S: Do you get a rash when you eat them or become short of breath or throw up?
>
> S: Is it an ASCII file?
> U: HOW_KNOW?
> S: Have you ever successfully listed it?
> **OR**
> S: Was it created by a FORTRAN program using FORMAT statements?

4. "Script"-like information relating events of interest with events that are perhaps more easily remembered by the user (Kolodner, 1981) — e.g.,

> S: Have you had a tetanus shot in the last two years?
> U: DON'T_RECALL
> S: Were you bitten by a dog or cat or some other animal in the last two years?
> U: Yes.
> S: Were you treated for the bite?
> U: Yes.
> S: Were you given a shot as part of the treatment?
> U: Yes.
> S: Then that was probably a tetanus shot.

5. teleological information concerning the reasons for making the request — i.e., the role of the requested information in the reasoning chain — e.g.,

> S: How tall are you?
> U: WHY?
> S: We would like to know whether your weight is right for your height.

6. the ability to do case analysis reasoning if the user can not answer a question. That is, the system can consider the effect on the outcome under the alternative assumption of each possible value or class of values. It is possible that after some analysis the system will discover that it does not make a difference which value is assumed. In the following example, the system could continue by carrying both possibilities forward (i.e., RhFactor ϵ {positive, negative}).

> S: What is your mother's Rh factor?
> U: DON'T_KNOW
> S: Could you ask her? We'll continue now without it.

It is clear to us that cooperative interactive problem-solving systems should be able to provide these capabilities for a user who either cannot or chooses not to answer its questions. We will continue to work on this problem here, and hope that work will be going on elsewhere as well.

5. Conclusion

In this paper we have illustrated additional capabilities that systems need if they are to move beyond straight factual question-answering to participating with their users in cooperative problem-solving interactions. Two of these we have discussed in more detail: getting systems to recognize and respond to users' misconceptions and enabling them to get from their users the information they need to help the users solve their problems. Without these capabilities, interactive problem-solving systems will never be more than laboratory toys.

6. References

Allen, J. Recognizing intentions from natural language utterances. In M. Brody (Ed.), *Computational models of discourse.* Cambridge, MA: MIT Press, 1982.

Ben Ari, M., Manna, Z., & Pneuli, A. The temporal logic of branching time. *Eighth Annual ACM Symposium on Principles of Programming Languages,* 1981.

Bobrow, D., Kaplan, R., Norman, D., Thompson, H. & Winograd, T. GUS, a frame driven dialog system. *Artificial Intelligence 8,* 1977.

Bobrow, R. J. *The RUS system* (Tech. Rep. 3878). Cambridge, MA: Bolt Beranek and Newman, Inc., 1978.

Cohen, P., Perrault, C. R. & Allen, J. *Beyond question-answering* (Tech. Rep. 4644). Cambridge, MA: Bolt Beranek & Newman Inc., 1981.

Davis, R. Interactive transfer of expertise. *Artificial Intelligence* 1979, *12*(2), 121–157.

Doyle, J. A truth maintenance system. *Artificial Intelligence* 1979, *12,* 231–272.

Engleman, C., Scarl, E. & Berg, C. Interactive frame instantiation. *Proceedings of the First National Conference on Artificial Intelligence (AAAI),* 1980.

Gazdar, G. *Pragmatics.* New York: Academic Press, 1975.

Grice, H. P. Logic and conversation. In P. Cole & J. L. Morgan (Eds.), *Syntax and Semantics.* New York: Academic Press, 1975.

Grosz, B. *The representation and use of focus in dialogue understanding* (Tech. Rep. 151). Menlo Park, CA: SRI International, 1977.

Hendrix, G., Sacerdoti, E., Sagalowicz, D. & Slocum, J. Developing a natural language interface to complex data. *ACM Transactions on Database Systems* 1978, *3*(2), 105–147.

Hirschberg, J. *Scalar quantity implicature: A strategy for processing scalar utterances* (Tech. Rep. MS–CIS–83–10). University of Pennsylvania, Computer and Information Science, May 1983.

Horn, L. Lexical incorporation, implicature, and the least effort principle. *Proceedings of the 14th Regional Meeting, Chicago Linguistic Society,* 1978.

Joshi, A. K. Mutual Beliefs in Question Answering Systems. In N. Smith (Ed.), *Mutual Belief.* New York: Academic Press, 1982.

Kaplan, J. Cooperative responses from a portable natural language database query system. In M. Brady (Ed.), *Computational Models of Discourse.* Cambridge, MA: MIT Press, 1982.

Kolodner, J. L. Organization and retrieval in a conceptual memory for events or CONS54, where are you? *Proceedings of the International Joint Conference on Artificial Intelligence,* 1981, 227–233.

Kwasny, S. & Sondheimer, N. Relaxation techniques for parsing ill-formed input. *American Journal of Computational Linguistics* 1981, *7*(2), 99–108.

Malhotra, A. *Design criteria for a knowledge-based English language system for management* (Tech. Rep. TR–146). Cambridge, MA: Project MAC. MIT Press, 1975.

Malhotra, A., Sheridan, P. *Experimental determination of design requirements for a program explanation system.* (Tech. Rep. RC 5831). Yorktown Heights, NY: IBM Research Center, 1976.

Mays, E. Failures in natural language systems: application to data base query systems. *Proceedings of the 1980 National Conference on Artificial Intelligence (AAAI),* August 1980.

Mays, E. Monitors as responses to questions: Determining competence. *Proceedings of the 1982 National Conference on Artificial Intelligence,* 1982.

Mays, E. A Modal temporal logic for reasoning about change. *Proceedings of the 1983 Association for Computational Linguistic Conference,* 1983.

McCoy, K. Correcting misconceptions: What to say. *CHI'83 Conference Human Factors in Computing Systems,* Cambridge MA, December, 1983.

McDermott, D., & Brooks, R. ARBY: Diagnosis with shallow causal model. *Proceedings of the National Conference on Artificial Intelligence,* (AAAI), 1982, 314–318.

McKeown, K. *Generating natural language text in response to questions about database structure.* Unpublished doctoral dissertation, University of Pennsylvania, 1982.

Pollack, M. *A framework for providing appropriate advice* (Tech. Rep. CIS 83–28). University of Pennsylvania, Computer & Information Science, October, 1983.

Pollack, M., Hirschberg, J., & Webber, B. User participation in the reasoning processes of expert systems. *Proceedings of the 1982 National Conference on Artificial Intelligence,*

(AAAI), 1982. (A longer version appears as Tech. Rep. CIS–82-9. University of Pennsylvania, Dept. of Computer and Information Science, July 1982.)

Robinson, J. DIAGRAM: A grammar for dialogues. *CACM,* 1982, *25*(1) 27–47.

Schmolze, J. G., & Brachman, R. J. Proceedings of the 1981 KL-One Workshop (Tech. Rep. 4842). Cambridge, MA: Bolt Beranek and Newman Inc., 1982. Also FLAIR TR-4, Fairchild Lab for AI Research.

Schuster, E. *Explaining and expounding* (Tech. Rep. MS–CIS–82–49). University of Pennsylvania, Computer & Information Science, 1982.

Schuster, E. Custom-made responses: Maintaining and updating the user model (Tech. Rep. MS–83–13). University of Pennsylvania, Computer & Information Science, September 1983.

Schank, R. C., & Slade, S. Advisory systems. In W. Reitman (Ed.), *Artificial intelligence applications for business.* Norwood, NJ: Ablex, 1984.

Shwartz, S. P. Natural language processing in the commercial world. In W. Reitman (Ed.), *Artificial intelligence applications for business.* Norwood, NJ: Ablex, 1984.

Sidner, C. L. Focusing in the comprehension of definite anaphora. In M. Brady (Ed.), *Computational Models of Discourse,* Cambridge, MA: MIT Press, 1982.

Sidner, C. L. & Israel, D. J. Recognizing intended meaning and speakers' plans. *Proceedings of the Seventh International Joint Conference on Artificial Intelligence,* 1981, 203–208.

Swartout, W. Explaining and justifying expert consulting programs. *Proceedings of the Seventh International Joint Conference on Artificial Intelligence,* 1981.

van Hahn, W., Hoeppner, W., Jameson, A., & Wahlster, W. The anatomy of the natural language dialogue system HAM–RPM. In L. Bolc (Ed.), *Natural Language based Computer Systems.* Munich: Hanser/Macmillan, 1980.

van Katwijk, A., van Nes, F., Bunt, H., Muller, H., & Leopold, F. *Naive subjects interacting with a conversing information system.* Eindhoven, Netherlands: IPO Annual Progress Report, 1979.

Wallis, J., & Shortliffe, E. Explanatory power for medical expert systems. *Methods Inf. Med,* 1982, *21*(3), 127–136.

Waltz, D. An English language question answering system for a large relational database. *CACM,* 1978, *21*(7), 526–539.

Webber, B. L. *A formal approach to discourse anaphora.* New York: Garland Press, 1978.

Webber, B. L. Pragmatics and database question answering. *Proceedings of the International Joint Conference on Artificial Intelligence,* 1983.

Wilensky, R. Talking to UNIX in English: An overview of UC. *Proceedings of the 1982 National Conference on Artificial Intelligence, AAAI,* 1982, 103–106.

Woods, W. A. Natural language communication with machines: An ongoing goal. In W. Reitman (Ed.), *Artificial Intelligence applications for business.* Norwood, NJ: Ablex, 1984.

Natural Language Processing in the Commercial World

Steven P. Shwartz
COGNITIVE SYSTEMS, INC.
234 CHURCH ST.
NEW HAVEN, CT 06510

Natural language processing technology gives the untrained user the capability of communicating with computers. This capability is a giant step in the evolution of user-friendly systems. Two classes of natural language systems, front ends and conversational advisory systems, are described, and the importance of a knowledge base is discussed.

People rarely have difficulty getting information into computers. The problems always arise when that information has to be retrieved. The development of the technology involved in storing information has consistently outpaced the development of the technology involved in retrieving information.

One of the major buzz-words in the data processing industry is "user-friendliness". The user-friendliness of a system can be judged by two criteria: the level of data processing skill required to operate the system; and the amount of time it takes to perform a given data processing task.

1. The Development of User-friendly Systems.

Historically, much of the progress in software technology can be viewed in terms of the increased user-friendliness of data processing systems. Users of the earliest computers were required to program in machine language. A portion of a sample program is shown in Figure 1.

Each line of code specifies one of the primitive instructions of the computer and the location of the data to which the instruction is to be applied. For example, a line in Figure 1 might specify that the contents of a particular memory address are to be moved into a particular register. In order to

.
.
.
37200146732
77326627215
42263530031
.
.
.

Figure 1. *Machine language code.*

perform a high-level operation such as adding two numbers, several machine instructions must be specified.

Machine language programmers require tremendous data processing skill. They must not only be familiar with basic programming constructs such as loops, procedures, and variables, but must also be familiar with the architecture of the computer itself. Further, the construction of each line of code requires looking up the numerical code for each instruction as well as a knowledge of the syntax necessary to specify the location information in the numeric code that represents the machine instruction.

The advent of assembler languages solved only the latter problem. Assembler languages offer the user a mnemonic means of specifying the numeric code corresponding to the machine instruction. That is, in order to move a number from a memory address into a register, the user need not look up and construct a numeric code, but can type something like

MOV 12345, R1.

Assembler languages certainly accelerate the coding process, but require essentially the same level of data processing skill as machine languages.

The next major advance in user-friendliness was the development of high-level languages such as FORTRAN (see Figure 2 below for an example).

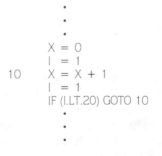

```
              .
              .
              .
          X = 0
          I = 1
   10     X = X + 1
          I = 1
          IF (I.LT.20) GOTO 10
              .
              .
              .
```

Figure 2. *Coding in FORTRAN.*

These languages considerably condensed the coding process by allowing the user to specify a computer program using far fewer instructions than would have been required to write the program in machine or assembler language. For example, adding two numbers typically requires only a single instruction. Further, the user is not required to be cognizant of the architecture of the machine being programmed. A program written in FORTRAN for one machine will typically run on another machine that supports FORTRAN with little or no modification. This eliminates the need for machine-specific knowledge on the part of the user. However, a high-level programmer still requires an understanding of programming concepts such as loops, variables, and procedures. While the number of high-level programmers greatly exceeds the number of machine or assembler language programmers, this level of data processing skill represents at least a year of training, and the number of people with this skill includes only a very small portion of the potential user population.

The 1970s produced two categories of systems designed to be accessible to users with less than a month of data processing training: non-procedural query languages and menu systems. A user of a query system tied to a personnel database need only type something like

SLCT EMPL JT = SLS DPT = MKT

in order to instruct the machine to select (SLCT) from the employee file (EMPL) all records in which the value of the JT (job-title) field is SLS (salesman) and the value of the DPT (department) field is MKT (marketing). Such query systems have a major advantage over programming languages in that the level of skill required to use such a system is much less.

The user of a query system does not need a knowledge of programming constructs such as loops, procedures, and variables. However, the query language user must learn how to select the appropriate mnemonics as well as the syntax by which these mnemonics are combined. Further, the user must understand the physical layout of the database (i.e., what information is contained in what files and how the information is divided into fields). Thus even query systems are far from being accessible to the novice user.

A minor enhancement to query systems is to make them English-like so that a user could type

SELECT EMPLOYEES JOB-TITLE = SALESMAN DEPARTMENT = MARKETING

in order to retrieve the same information as in the previous example. The difference, of course, is that in this case the mnemonics are much easier to remember because they are English words. However, the level of data processing skill on the part of the user is the same as in a non-English-like query language. The user must still learn what words are correct, the syntax, and the database structure. For example, the user could not simply say

LIST ALL THE SALESMEN IN THE MARKETING DEPARTMENT.

Menu systems, in contrast to query languages, require almost no training, but are more tedious than query languages and will very often fail to give the user the information desired. The operating procedure in a menu system is very simple. In a typical menu system, the user is presented with a series of numbered choices, the user indicates one choice by entering the number associated with that choice, and the system presents another series of choices to the user designed to further delineate exactly what information the user desires. A sample screen from a shop-at-home menu system is shown in Figure 3.

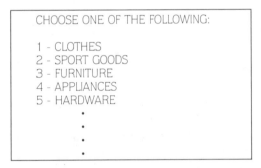

CHOOSE ONE OF THE FOLLOWING:

1 - CLOTHES
2 - SPORT GOODS
3 - FURNITURE
4 - APPLIANCES
5 - HARDWARE
·
·
·
·

Figure 3. A sample menu for a shop-at-home system.

The problems with menu systems fall into two categories: First, the user often must page through on the order of a dozen screens in order to get the desired information. This becomes extremely frustrating to the experienced user, particularly when the user is making similar retrieval requests time after time. A more serious problem is that all too often a user is presented with a menu in which the correct choice is not clear. Suppose, for example, that the user of the system in Figure 3 wanted to buy a clock-radio. Which choice is correct, FURNITURE or APPLIANCES? The problem here is not one of a poorly designed set of choices. If enough choices were included to specify choices at the level of detail of "clock-radio", the number of screens (or at least the size of the screens) would multiply quickly. The designer of a menu system always faces a tradeoff between the amount of information that will be accessible to the user and the number of screens the user will need to traverse in order to get even the simplest information.

2. Natural Language Front End Systems.

The technology of natural language processing provides the vehicle by which truly naive users can gain access to the information stored in a computer. A system with a natural language interface is one in which the user communicates with the system in English, French, German, or any other natural language, as opposed to artificial languages such as FORTRAN, COBOL, or a query language. The key feature of a natural language system is that, although the user may be restricted to a certain subject area in interacting with the system, the user is not restricted to a subset of English or of some other natural language. Further, the user is not required to have an understanding of the physical layout of the data within the system (in contrast to query and menu systems, as discussed above).

The obvious application of natural language technology is to provide natural language access to information in a database. In order for this type of system, termed a natural language front end, to be used by people with no data processing training whatsoever, the system must simulate the behavior of a programmer or technician whose job it is to retrieve information from a database in a traditional data processing facility. In order for a natural language query system to meet this programmer/technician criterion, it must

1. Process the full gamut of syntactic constructs—including ungrammatical requests.
2. Process requests varying in complexity.
3. Process incomplete and/or ambiguous requests by engaging the user in an interactive dialogue.
4. Process a request with respect to its discourse context.

2.1. EXPLORER: An Example

An example of such a natural language front end is the EXPLORER system developed by Cognitive Systems Inc. for a major oil company. EXPLORER is a natural language front end to a database containing approximately 175 fields of information on over one million oil wells, as well as information on most of the geological formations in North America. Those exploring for oil (typically geologists and geophysicists) use the database primarily to generate maps of particular geographic regions. These maps show contour information (e.g., the depth contours) of the formations within the region as well as the characteristics of the wells within the region. Prior to development of the natural language front end, a mapping system was available for the purpose of creating such maps. The mapping system had an English-like query system interface (see above) that required a few days of training to learn. However, the oil company had a great deal of difficulty getting its people to use the system, for a variety of reasons. Some could not successfully complete the training. More often, the system would be learned successfully, but forgotten due to intermittent use. Equally often, individuals would refuse any form of data processing training. As a result, the oil company hired a large team of technicians to interface between the geologists and geophysicists and the query system. The geologist would tell the technician what type of map was desired and the technician would write the English-like query.

The EXPLORER system was designed to simulate this technician. A sample dialogue with EXPLORER is shown below in Figure 4.

EXPLORER   Version 05   10/15/83

PLEASE ENTER A MAP REQUEST:

*Please give me an isopach map from the smackover to the
*eagleford in Hemphill County, Texas with a contour interval
*of 100 feet and a scle of 1″ = 2000′.

By SCLE, do you mean SCALE (Y or N)? y

User requests an ISOPACH map
output medium: PAPER
output device: PHOTODOT
county: HEMPHILL, TEXAS
from SMACKOVER to EAGLEFORD
ci: 100 feet
scale: 1 inch = 2000 feet

Figure 4. *Sample request to EXPLORER and the resulting paraphrase.*

The initial user input was a request for an isopach map, which is a map showing the thickness of the earth between two geologic formations, in this case the smackover and the eagleford. The user has also specified a map region, a contour interval (i.e., the distances between contour circles on the map) and a scale for the map. The user misspelled scale in this example, and the system responded with a correct spelling. This capability is extremely important in a natural language system, because people unused to computer keyboards are extremely prone to spelling errors.

The system then responded with a paraphrase of the user's request, including a couple of default output specifications. The dialogue then continues:

Would you like to make any changes or
    additions to your request? (Y or N) *y

Please input changes or additions.
*I wanted the map on film not paper.

User requests an ISOPACH map
output medium: FILM
output device: PHOTODOT
county: HEMPHILL, TEXAS
from SMACKOVER to EAGLEFORD
ci = 100 feet
scale of 1 inch = 2000 feet

Would you like to make any changes or
    additions to your request? (Y or N) *y

Please input changes or additions.
*Give me only wildcats that are not deeper than 2000'.

User requests an ISOPACH map
output medium: FILM
output device: PHOTODOT
county: HEMPHILL, TEXAS
from SMACKOVER TO EAGLEFORD
filters: WILDCAT WELLS ONLY
        DRILLING DEPTH < 2000 feet
ci = 100 feet
scale of 1 inch = 2000 feet

Would you like to make any changes or
    additions to your request? (Y or N) *y

Please input changes or additions.
*Show only wells with a well show deeper than 1500'

User requests an ISOPACH map
output medium: FILM
output device: PHOTODOT
county: HEMPHILL, TEXAS
from SMACKOVER to EAGLEFORD
filters: WILDCAT WELLS ONLY
        DRILLING DEPTH <= 2000 FEET
        WELL SHOW OF OIL > 1500 feet
ci = 100 feet
scale of 1 inch = 2000 feet

Would you like to make any changes or
    additions to your request? (Y or N) *n

Figure 5. *Interactive dialogue from the EXPLORER program.*

After displaying a paraphrase of the user's request, the system goes into
a verification loop, offering the user the capability of verifying or modifying
the request before the map is actually produced. This is essentially the

same service the human technician provides. That is, after the geologist tells the human technician what type of map is desired, the technician typically interacts with the geologist to verify and/or clarify his understanding of what the geologist wants. In the above example, the user is not happy with one of the output default chosen by the system, and requests that the map be output on film rather than on paper.

Then in the two subsequent interactions, the user specifies that only certain wells should be displayed on the map. If no such specifications were made, all wells in the area would be displayed. In this example, the user asks only to see wildcat wells drilled to a certain depth. In this example, the system must infer that "deeper than" refers to the total drilling depth of the well and must understand how to apply negation to a drilling depth specification.

Another capability of the EXPLORER system is the ability to understand map requests involving contextual reference. For example, if the user had just made the map in the previous example, and now wanted a map with the same parameters, but with different contour information, the user might type:

> PLEASE ENTER A MAP REQUEST:
>
> *Please give a structure map of the smackover that is like the last map.

Figure 6. A request making a contextual reference.

One value of a natural language system is that it can accommodate users with a wide range of experience with the system. That is, first-time users will typically not know what a complete map specification entails, while experienced users will typically want the capability of typing a concise request rather than an entire English sentence. EXPLORER offers both capabilities. For example, a first-time user wanting a map of a particular formation might type simply

> PLEASE ENTER A MAP REQUEST:
>
> *Map the smackover.

Figure 7. A request from a naive user.

It is then up to the system to guide the user by the hand in specifying enough information to make a map. This is illustrated in Figure 8.

> A map region was not specified.
>
> Do you wish to specify the map region by county(C) or by geographic coordinates(G)? *c
>
> Please enter a county (or counties) by name or number.
> *bibb county

```
BIBB County is in the following states:
  ALABAMA GEORGIA
Please enter the appropriate state or a new county specification.
*ala

Please specify contour interval.
*100'

Please specify scale.
*1 to 2000

User requests a STRUCTURE map
output medium: PAPER
output device: PHOTODOT
county: BIBB, ALABAMA
top: SMACKOVER
ci = 100 feet
scale of 1 inch = 2000 feet
```

Figure 8. *Interactive dialogue with a naive user.*

Here the system has led the user by the hand to a request specification containing enough information to make a map.

At the other extreme the experienced user wants to type something like

```
PLEASE ENTER A MAP REQUEST:

*smackover bibb georgia ci 100 scale 2000.

    User requests a STRUCTURE map
    output medium: PAPER
    output device: PHOTODOT
    county: BIBB, GEORGIA
    top: SMACKOVER
    ci = 100 feet
    scale of 1 inch = 2000 feet
```

Figure 9. A *request from an experienced user*

This type of telegraphic request, containing poor syntax or even lacking syntax altogether, is acceptable to the EXPLORER system. The criterion of a natural language front end should be only that the user input be comprehensible to the technician that the system is simulating.

2.2. How EXPLORER *works*

EXPLORER embodies natural language analysis techniques developed at the Yale Artificial Intelligence Project, the most important of which is a conceptual analyzer (Dyer, 1981; 1982) that uses expectation-driven parsing techniques (Birnbaum & Selfridge 1979; Dyer, 1981, 1982; Gershman 1979; Riesbeck 1975; Riesbeck & Schank 1976; Schank and Birnbaum 1980). The output of the conceptual analyzer is a language-free meaning representation (Schank & Abelson 1977), and all question-answering and memory search techniques are derived from the work of Lehnert (1978) and Dyer (1982).

The philosophy behind EXPLORER is based on the notion that the language understanding process utilizes syntactic, semantic, pragmatic and contextual cues in an integrated fashion. The EXPLORER system does *not* perform distinct syntactic, semantic, pragmatic and contextual analyses in a linear sequence, but rather attempts to utilize each type of analysis in conjunction to determine the meaning of an input string. The input string is analyzed word-by-word, and an attempt is made to build a conceptual representation of the input string that keeps pace with the word-by-word analysis. Syntactic, semantic, pragmatic, and contextual information are all treated as cues that can aid this process. As the conceptual representation is constructed, expectations regarding the remainder of the input string are set up. The expectations can be for syntactic constructions, conceptual content, or likely inferences. When ambiguous words are encountered, expectations that will resolve these ambiguities are generated. (For more detail on the parsing process, see Birnbaum & Selfridge 1979; Dyer, 1981, 1982; Gershman 1979; Riesbeck 1975; Riesbeck & Schank 1976; Schank & Birnbaum 1980.)

3. Why a Knowledge Base Is So Important.

One issue that comes up time and time again concerns the transportability of the natural language capability resident in the EXPLORER to domains other than oil exploration. Unfortunately, the power of the EXPLORER system is due mainly to the system's knowledge concerning oil exploration. While the natural language capability is a general one, approximately ½ to 2 years are required to code the domain-specific knowledge inherent in a knowledge-based system.

In a sense, the phrase natural language processing is misleading. The task of building a natural language system is more of a knowledge engineering task than a linguistic one. One obvious requirement for knowledge is in the system's processing of incomplete requests. That is, the code involved in guiding the user by the hand is clearly specific to the task of generating map requests.

It is less obvious, however, that knowledge of oil exploration is crucial to actually understanding the user's natural language inputs. For example, in Figure 4, the user requests only wells "not deeper than 2000′". In order to understand this input, it is necessary for the system to infer that "deeper than" refers to the total drilling depth of the well. This inference, however, is context-dependent and knowledge is needed to determine where the inference is appropriate. That is, it would not suffice to simply identify the phrase "deeper than" as synonymous with "total drilling depth". If this were done, then the request in Figure 4, ". . . wells showing oil deeper than 1500'" would be synonymous with ". . . wells showing oil total drilling depth 1500'" and would lead to a misinterpretation.

Another example is the application of the negative in "wells not deeper than 2000'" (Figure 4). Correct application of the negative requires an understanding of the notion of a drilling depth. In fact, there is a wide range of linguistic phenomena that represent unsolved problems for non-knowledge-based domain-specific systems. Examples include the resolution of anaphora, quantification, time specifications, accessing multiple files, and queries involving aggregation (see Shwartz, 1982 for further details).

4. Domain-independent Natural Language Systems.

There are systems that claim to have natural language capability that is domain-independent. That is, they will work in any subject area without the need for domain-specific knowledge engineering. These systems essentially circumvent the need for knowledge by examining the physical structure of the database. That is, these systems can process a natural language request such as

LIST ALL THE SALESMEN IN THE MARKETING DEPARTMENT.

by recognizing that SALESMEN is a possible value of the JOB-TITLE field in the EMPLOYEE file and that MARKETING is a possible value of the DE-PARTMENT field. Using this device, such systems can understand certain natural language requests, but cannot achieve the level of understanding of a knowledge-based system.

To get a feeling for the level of understanding of such a system, imagine having a database in which the English field names were replaced by nonsense names. If a person were given such a database and had the ability to examine fields and files it would be possible to answer such questions as

LIST ALL THE GORKLES IN THE MIZBING.

provided that GORKLES and MIZBING were identifiable file names, field names, or field values. However, suppose the request were

LIST OUR 3 BEST GORKLES.

There is no way of answering this question without knowing whether GOR-KLES are salesmen, stocks, or baseball pitchers. That is, a person could answer this query by accessing world knowledge concerning business, securities, and baseball, respectively. A knowledge-based system can do the same thing. A domain-independent system, however, will behave at best like a person who knows nothing about salesmen, stocks, baseball, or gorkles.

5. Conversational Advisory Systems.

As described above, a natural language front end is a simulation of a human technician. As such it has a relatively low knowledge level. It is possible to upgrade the degree of knowledge in the system and to simulate a person with a higher level of skill who can give advice, such as a travel agent or a financial advisor. We call this type of system a conversational advisory system.

We use the term "advisory" to distinguish this type of system from what are traditionally termed "expert systems". Expert systems typically embody the knowledge of someone with a relatively unique expertise. For example, the dipmeter advisor (Gershman, 1982) was designed to interpret a particular type of geological test that even very few geologists were skillfull at interpreting. A conversational advisory system is designed to simulate a person with a more mundane skill, albeit a highly proficient practitioner of that skill.

5.1. WILL-WRITER: An Example

A prototype of a conversational advisory system for an automatic will-writer was developed at Cognitive Systems by Michael Dyer. Figure 10 contains a sample run of that system.

> Welcome to The Cognitive Systems
> Automatic Will Writer and Estate Planner.
>
> .
> .
> .
> .
>
> How many children do you have?
> >two
>
> Please enter the first name and then the age (in years) of your ELDEST child.
> >Hanna, age 10
>
> Please enter the first name and then the age (in years) of your YOUNGEST child.
> >Joshua 6

Figure 10. *Initial data-gathering dialogue with* WILL-WRITER.

Up to this point, the system has functioned like a traditional data-gathering system, asking the user questions and receiving straightforward answers. A functional, but not particularly important feature of the system is that the user's answers are free-format. At this point, the system will deviate from this pattern and give the user some advice (Figure 11).

I see that you have young children. The best way to provide for them (in case you and your wife should die in an accident) is to set up a TESTAMENTARY TRUST in your will.

Please enter the legal name of a close friend or relative (not counting your spouse) who would be willing and able to manage your children's share of your estate in the event of your death. (First Middle Last)

Figure 11. *Advice given by* WILL-WRITER.

This advice-giving capability is available with traditional data processing technology. The uniqueness of a conversational advisory system is that the user can interact with the system and discuss the advice given. A traditional advisory system is like a black box—one must either take or leave the advice. In Figure 12, the dialogue terminates with the system asking the user to enter a name. In a traditional system, a name would be the only acceptable response. In a conversational advisory system, the user has the option of discussing the advice given.

>What does "testamentary" mean?

There are two types of trusts: LIVING TRUSTS and TESTAMENTARY TRUSTS. A LIVING TRUST takes effect while the creator of the trust is still alive. A TESTAMENTARY TRUST only comes into existence when the trust donor is dead.

So the trust for your children is called a TESTAMENTARY TRUST because it is specified in your will and only becomes active once you are deceased.

Figure 12. *Conversational dialogue clarifying advice.*

6. Conclusions

Natural language processing technology gives the untrained user access to computers for the first time. However, not all natural language systems are created equal. Natural language systems lacking a knowledge base cannot understand anywhere near as wide a range of information retrieval requests as can knowledge-based systems. More importantly, only knowledge-based systems can be extended in conversational advisory systems— a class of natural language systems which may well ultimately provide the most important user services.

7. References

Birnbaum, L., & Selfridge, M. Problems in conceptual analysis of natural language. In R. Schank & C. Riesbeck (Eds.), *Inside computer understanding.* Hillsdale, NJ: Lawrence Erlbaum Assosciates, 1981.

Dyer, M. G. Integration, unification, reconstruction, modification: An eternal parsing braid. In *Proceedings of the Seventh International Joint Conference on Artificial Intelligence,* August 1981.

Dyer, M. *In-depth understanding: A computer model of integrated processing for narrative comprehension* (Research Report #219). Yale University, Computer Science Department, 1982.

Gershman, A. V. *Knowledge-based parsing* (Research Report #156). Yale University, Computer Science Department, 1979.

Gershman, A. V. Building a geological expert system for dipmeter interpretation. *Proceedings of the European Artificial Intelligence Conference.* July 1982.

Lehnert, W. *The process of question answering.* Hillsdale, NJ: Lawrence Erlbaum Associates, 1978.

Riesbeck, C. Conceptual analysis. In R. C. Schank (Ed.), *Conceptual information processing.* Amsterdam: North Holland, 1975.

Riesbeck, C. K. & Schank, R. C. *Comprehension by computer: Expectation-based analysis of sentences in context* (Research Report #78). Yale University, Computer Science Department. Also in W. J. M. Levelt & G. B. Flores d'Arcais (Eds.), *Studies in the perception of language.* Chichester, England: John Wiley & Sons, Ltd., 1979.

Schank, R. C. & Abelson, R. *Scripts, plans, goals and understanding.* Hillsdale, NJ: Lawrence Erlbaum Associates, 1977.

Schank, R. C. & Birnbaum, L. *Memory, meaning and syntax* (Technical Report #189). Yale University, Computer Science Department, 1980.

Shwartz, S. P. Problems with domain-independent natural language database access systems. *Proceedings of the Association for Computational Linguistics.* June 1982.

Advisory Systems*

Roger C. Schank and Stephen Slade
DEPARTMENT OF COMPUTER SCIENCE
YALE UNIVERSITY
P.O. BOX 2158
NEW HAVEN, CT 06520

We discuss the general model of expert programs to offer advice. These programs are proposed as an application of current research in modeling human memory in domains requiring decision making. The main feature that distinguishes this memory-based advisory system from the typical rule-based expert system is the representation of knowledge as experience, rather than rules. This approach is designed to allow programs to learn more about domains as new situations are encountered; to provide a mechanism for a deeper level of analysis if failures are encountered; and to facilitate the inference of the goals of the user and to accommodate those goals in the advice that is given.

1. The Problem: Decision Making

A fundamental task of a business man is to make decisions.

- What products should I develop?
- How should I raise capital?
- What should I invest in?
- Do I need to hire more programmers?
- Should we buy 50 PC's or a new mainframe?
- Will AI improve my company's productivity?
- Are we spending enough on R&D?
- Are there new markets we should explore?

---

* The research described here was done at the Yale Artificial Intelligence Project and is funded in part by the Air Force Office of Scientific Research under contract F49620–82–K–0010, and the Defense Advanced Research Projects Agency and monitored by the Office of Naval Research under contract N00014–75–C–1111.

People must recognize when they need to make a decision, what the possible outcomes are, what the likelihoods of those outcomes are, what information is relevant to this decision, and how to get that information.

Clearly, decision making is a very complex activity. It involves lots of different kinds of knowledge and planning behavior. It is also an intrinsic cognitive activity. That is, decision making is psychological in nature. There have been attempts in economics to develop rational models of decision making, but these have failed to capture the rich complexity of human decision making.

AI has a fundamental interest in modeling decision making in general. A lot of AI work has been done in various constrained decision-making domains, such as game playing (Samuel, 1963), and medical diagnosis (Shortliffe 1976). The area of *expert systems* can be seen automating a part of the decision making process for particular domains.

In business, the problems are often very complex. Information technology has developed extremely rapidly over the past several decades. Telecommunications provides instant access to remote locations. Computers store and retrieve tremendous amounts of data. The problem for business is to determine what of that data is important information.

2. The Solution: Advisory Systems

When a business has a complex decision or problem, it often turns to experts for advice. These experts have specific knowledge and experience in the domain. They are aware of what alternatives may be available, what the probability of success is, and what costs the business may incur. Lawyers or marketing specialists or environmental engineers or public relations firms all have particular expertise that informs the advice they can give their clients. The more familiar the expert is with the specific client or industry or problem, the better (and generally, more expensive) the advice. If the expert knows a lot about the current situation, and previous similar situations, he is better prepared to help the business make a decision.

Current AI expert systems are an attempt to model the knowledge and ability of such experts. Typically, these programs are done in conjunction with a human expert in the field. This person works with a programmer to develop a set of conditional rules for solving problems in the domain.

One should note one very important fact about expert systems and advisory systems in general: they make mistakes. Just like human experts, they are not 100% reliable. Problems which have absolute solutions do not require expert systems. Thus, there are no expert systems for arithmetic or sorting. There are expert systems for oil exploration, computer component configuration, and equipment failure repair. The goal in most of the expert systems is to achieve a success rate on par or better than the human expert.

The heart of expert systems is the set of rules. These rules take the form: IF <condition> THEN <action>. Typically, that is the primary knowledge base. The program uses these rules and only these rules in solving problems. If the program is unable to solve a problem, or reaches a wrong conclusion, it is up to the programmer to diagnose the error and correct the rule or add another rule.

Furthermore, most AI expert systems are not able to learn from experience. After a program has encountered a certain situation and observed what the outcome was, the program should be able to remember that episode and use that knowledge when a subsequent similar situation arises. This type of learning and adapting is a difficult problem and is not typically available in rule-based expert systems. That is, a rule-based system does not generally change its rules on the basis of a problem submitted to it. Only the programmer can alter the rules.

The rules themselves in the expert system are not easy to come by. The knowledge acquisition task is very time-consuming and inexact. Formulating rules is a difficult job. A program that could develop its own rules — possibly by modifying or specializing some initial set of rules — would be a great boon. Since the core of expert systems is the knowledge itself, we should find ways of having the system acquire and refine its own knowledge.

One final feature one would want in advisory expert systems is for the program not only to understand the problem, but understand it in terms specific to the client/user. It is critical that the program examine a problem in the particular context of the present client. The program has to know about the client's goals and needs; what specific ramifications each alternative may have for the client; how the decision might affect the client's employees, owners, competitors, customers, and suppliers; and what priorities the clients places on these possible effects.

The importance of understanding a client's goals can be seen in the domain of investment advice. Not only does the program have to understand how to modify its plans for a risk-averse, conservative client versus a risk-taking client, the program also has to be able to recognize one from the other. For example, a single 30-year-old male earning $25,000 per year and a 60-year-old widow with four children earning $25,000 per year most likely have very different investment goals. Furthermore, while it would be possible to incorporate that type of knowledge about young men and old women into a set of rules, one should recognize that there are causal explanations involved that are quite general and have very little to do with the specific domain of investing. We have knowledge about life expectancy and responsibility for taking care of dependents and the cost of living that touches on the analysis. A robust financial advisor needs a great deal of world knowledge.

We thus propose three basic features for advisory systems of the future. These computer programs must:

1. *Learn from experience.* The knowledge that a computer requires will come from exposure to new situations which will be integrated into the memory of the computer.
2. *Adapt to failures.* When the program makes mistakes or encounters something unexpected, it must be able to react and adjust, instead of simply bombing out.
3. *Infer the user's goals.* The program must be able to adapt to new users as well as new problems. The specific needs of the user must be taken into account.

3. Learning From Experience

How can a computer get all that knowledge? How should that knowledge be organized in a computer? How can a computer learn?

In our work over the years in natural language understanding, we have focused on modeling human cognitive behavior. There may be other ways to get computers to communicate using natural language; however we have always believed that the psychological method provided the most productive approach. We produced computer programs that could read stories and then perform certain tasks, such as question-answering, paraphrase or translation, which demonstrated that the computer had in some sense understood the meaning of the text. These early programs, MARGIE (Schank, 1975), SAM (Cullingford, 1978), PAM (Wilensky, 1978), POLITICS (Carbonell, 1979), and FRUMP (DeJong, 1979), were based on psychological theories of language comprehension and cognitive behavior. FRUMP, for example, was a program which could read newspaper stories directly from a United Press International wire and produce summaries of the stories in several foreign languages. FRUMP was a model of a person skimming through news stories, and would only notice those salient details in the 50 or 60 domains for which it had knowledge. There was one significant problem with FRUMP, and the other programs for that matter, from the psychological perspective: there was no memory. That is, these programs could (and did) read the same story over and over again and not recognize that they had read it before. Clearly, a person would not behave in that manner.

The current generation of expert systems are beset with the same deficiency. They have no memory. If the same case is presented to an expert system repeatedly, the system will perform in exactly the same manner in each case. So what?, you may say. You want the system to be reliable and what does psychological validity matter as long as the system works? Why bother building a memory of episodes if it's not going to make the program run any better?

However, we maintain that a memory based system will prove better and more reliable than its rule-based counterpart. The reason lies in the origin of the rules themselves. Think, if you will, of the human expert. What is the nature of his expertise? The expert has rules, but where did those rules come from? An expert, fundamentally, has *experience.* It is experience which forms the core of expertise. When an expert encounters a new situation, he may be reminded of a previous similar episode whose outcome is known to him. He can then reason about the new situation in terms of the previous one and make predictions about alternative actions based on this experience.

When an expert has encountered numerous experiences that are greatly similar, he may no longer have specific remindings, but will have developed a more general representation or distillation of these experiences. That is, the expert derives a rule. It is these rules that form the knowledge framework for the rule-based expert systems. However, there is an important difference between the rule in the computer program and the expert's rule. The latter rule is malleable and the former is not. If the expert gets a new situation that does not quite match the rule, the expert can still draw on his experience that formed the basis for the rule to analyze the novel situation.

As we mentioned above, our work in story understanding lead us to be concerned about modeling memory and reminding. A subsequent story understanding program, IPP (Lebowitz, 1980; Schank, Lebowitz, & Birnbaum, 1978), was able to remember stories that it had read and use that knowledge to handle later stories. IPP read hundreds of newspaper stories and was able to draw several dozen meaningful generalizations from those stories. As an example of a computer program which learns from experience, IPP serves as a prototype for the memory-based expert advisory systems we are discussing here.

IPP's particular area of experience was stories dealing with terrorism. It started off with general knowledge about terrorist acts, such as kidnappings, bombings, and shootings, and was able to add to its knowledge by reading stories — not by having a programmer simply add more rules.

The following is an annotated example of IPP reading several stories about terrorist acts in Northern Ireland. MOPs are Memory Organization Packets, which are explained in Schank (1982). MOPs are used in IPP to establish a hierarchy of experiences in memory. In this example, IPP reads two stories and notices that both take place in Northern Ireland, that the victim is an establishment, authority figure, and that the perpetrator is a member of the IRA. This fact becomes a general rule which is applied in a third story which deals with a shooting of a policeman in Northern Ireland. Though it is not stated in the story, IPP infers that the gunman is a member of the IRA. Admittedly, this may not be correct, but it is the type of inference which a person would be inclined to make.

Comments will appear throughout the log with a "~" in front of them.

@<demo>ipp

IPP created 16-Jan-81 15:26:05, ready 26-Jan-81 15:44:03

*(parse xx1)

Story: XX1 (4 12 79) NORTHERN-IRELAND NONE

(IRISH REPUBLICAN ARMY GUERRILLAS AMBUSHED A MILITARY PATROL IN WEST BELFAST YESTERDAY KILLING ONE BRITISH SOLDIER AND BADLY WOUNDING ANOTHER ARMY HEADQUARTERS REPORTED)

Processing:

IRISH REPUBLICAN ARMY
                    : Phrase
IRA   : Token refiner—save and skip
GUERRILLAS   : Interesting token—GUERRILLAS OF THE IRA

Instantiated I-TERRORISM structure

Predictions—I-TERRORISM-SYN-FINDER I-TERRORISM-SUB-STRUCTURE
             INFER-I-TERRORISM-SUB-STRUCTURES REDUNDANT-SCRIPT-
             WORDS
             DEFAULT-ORGANIZATION DEFAULT-VICTIM-TYPE COUNTER-
             MEASURES

~ IPP goes to its long-term memory whenever it has added new MOPs or
~ new features to its representation of the story.

>>> Beginning memory update . . .

New features: EVO (XX1) (I-TERRORISM)
ACTOR   ORG   IRA
                 POL-POS   BAD-GUY
                 AFFILIATION CATHOLIC
LOCATION   AREA   WESTERN-EUROPE
                 NATION   *NORTHERN-IRELAND*

Best existing S-MOP(s)—
I-TERRORISM

~ When it does a feature analysis and searches long-term memory, it chooses
~ best spec-MOP which it has in memory for the story. Since no spec-MOPs
~ are in memory right now, it chooses the default S-MOP (or in this case,
~ I-MOP).

>>> Memory update complete

AMBUSHED   : Memory sub-structure word

Instantiated $AMBUSH structure

Predictions—REDUNDANT-SCRIPT-WORDS $AMBUSH-ROLE-FINDER END-ROLE-
             FINDER

Instantiated S-ATTACK-PERSON structure

Predictions—DEFAULT-METHOD S-ATTACK-PERSON-SUB-STRUCTURE
              INFER-S-ATTACK-PERSON-SUB-STRUCTURES REDUNDANT-SCRIPT-
              WORDS DEFAULT-KILL

Prediction confirmed—S ATTACK-PERSON-SUB-STRUCTURES

>>> Beginning memory update . . .

New features: EV2 (XX1) (S-ATTACK-PERSON)
ACTOR   ORG   IRA
                POL-POS   BAD-GUY
                AFFILIATION   CATHOLIC
METHODS   AU $AMBUSH
LOCATION   AREA   WESTERN-EUROPE
                NATION   *NORTHERN-IRELAND*
I-MOP   I-MOP   I-TERRORISM

    Best existing S-MOP(s)—
S-ATTACK-PERSON

New features: EVO (XX1) (I-TERRORISM)
INSTANCES   AU   S-ATTACK-PERSON

    Best existing S-MOP(s)—
I-TERRORISM

    >>> Memory update complete

    A   : Function word—Token refiner—save and skip
    MILITARY   : Token refiner—save and skip
    PATROL   : Normal token—PATROL

Prediction confirmed—$AMBUSH-ROLE-FINDER(VICTIM)
Inferring victim nationality: *NORTHERN-IRELAND*

>>> Beginning memory update . . .

New features: EV2 (XX1) (S-ATTACK-PERSON)
VICTIM   POLITICS   MILITARY
                ROLE   AUTHORITY
                POL-POS   ESTAB

    Best existing S-MOP(s)—
    S-ATTACK-PERSON

New features: EVO (XX1) (I-TERRORISM)
VICTIM   POLITICS   MILITARY
                ROLE   AUTHORITY
                POL-POS   ESTAB

    Best existing S-MOP(s)—
    I-TERRORISM

    >>> Memory update complete

    IN   : Function word—preposition—
    WEST   : Token refiner—save and skip
    BELFAST   : Normal token—BELFAST
    YESTERDAY   : Normal token—YESTERDAY
    KILLING   : Word satisfies prediction

Prediction confirmed—CAUSE-DEATH-SUB-STRUCTURE

Instantiated CAUSE-DEATH structure

Predictions—REDUNDANT-SCRIPT-WORDS INFER-REASON-FOR-DEATH
             FIND-REASON-FOR-DEATH IMPORTANT-VICTIM

>>> Beginning memory update . . .

New features: EV2 (XX1) (S-ATTACK-PERSON)
RESULTS   AU   HURT-PERSON
               HEALTH   −10

Best existing S-MOP(s)
S-ATTACK-PERSON

>>> Memory update complete

ONE (ONE 2)   : Token refiner—save and skip
BRITISH (*ENGLAND*)
                   : Token refiner—save and skip
SOLDIER   : Normal token—SOLDIER
AND   : Function word—conjunction—save and skip
BADLY   : Token refiner—save and skip
WOUNDING   : Word satisfies prediction

Prediction confirmed—CAUSE-WOUND-SUB-STRUCTURE

Instantiated CAUSE-WOUND structure

Predictions—CAUSE-WOUND-SYN-FINDER END-ROLE-FINDER REDUNDANT-
             SCRIPT-WORDS IMPORTANT-VICTIM CAUSE-WOUND-ROLE-FINDER
             END-ROLE-FINDER

>>> Beginning memory update . . .

New features: EV2 (XX1) (S-ATTACK-PERSON)
RESULTS   HEALTH   −5

Best existing S-MOP(s)—
S-ATTACK-PERSON

>>> Memory update complete

ANOTHER (ANOTHER1)
                   : Token refiner—save and skip

Predictions—FIND-HISTORICAL-MODE

ARMY   : Token refiner—save and skip
HEADQUARTERS   : Normal token—HEADQUARTERS
REPORTED   : Dull verb—skipped

Final memory update/default processing

>>> Beginning final memory incorporation . . .

Feature analysis: EV2 (S-ATTACK-PERSON)
RESULTS   HEALTH   −5
               AU   HURT-PERSON
               HEALTH   −10
VICTIM   POLITICS   MILITARY
               ROLE   AUTHORITY
               POL-POS   ESTAB

ACTOR	ORG	IRA
	POL-POS	BAD-GUY
	AFFILIATION	CATHOLIC
METHODS	AU	$AMBUSH
LOCATION	AREA	WESTERN-EUROPE
	NATION	*NORTHERN-IRELAND*
I-MOP	I-MOP	I-TERRORISM

~ When the story is done, it is remembered in long-term memory with a pointer
~ to the closest spec-MOP, which it chose during its processing of the story,
~ (in this case, it had to choose the default S-ATTACK-PERSON, since memory
~ had no spec-MOPs), and a list of features which it has extracted from the
~ story which are not included in the spec-MOP.

Indexing EV2 (XX1) as variant of S-ATTACK-PERSON

Feature analysis: EV0 (I-TERRORISM)

VICTIM	POLITICS	MILITARY
	ROLE	AUTHORITY
	POL-POS	ESTAB
INSTANCES	AU	S-ATTACK-PERSON
ACTOR	ORG	IRA
	POL-POS	BAD-GUY
	AFFILIATION	CATHOLIC
LOCATION	AREA	WESTERN-EUROPE
	NATION	*NORTHERN-IRELAND*

Indexing EV0 (XX1) as variant of I-TERRORISM

>>> Memory incorporation complete

Story Representation:

```
** MAIN EVENT **
EV0 =
  MEM-NAME     I-TERRORISM
  ACTOR        GUERRILLAS OF THE IRA
  VICTIM       IRISH MILITARY PATROL
  INSTANCES
  EV2 =
    MEM-NAME     S-ATTACK-PERSON
    ACTOR        GUERRILLAS OF THE IRA
    VICTIM       IRISH MILITARY PATROL
    METHODS
    EV1 =
      MEM-NAME     $AMBUSH
      ACTOR        GUERRILLAS OF THE IRA
      VICTIM       IRISH MILITARY PATROL
    RESULTS
    EV3 =
      MEM-NAME     CAUSE-DEATH
      ACTOR        GUERRILLAS OF THE IRA
      VICTIM       IRISH MILITARY PATROL
      HEALTH       −10
    EV4 =
      MEMNAME      CAUSE-WOUND
```

```
        ACTOR              GUERRILLAS OF THE IRA
        HEALTH          −5
TIME        YESTERDAY
```

5881 msec CPU (0 msec GC), 43000 msec clock, 7438 conses
NIL

*(parse xx2)

Story: XX2 (11 11 79) NORTHERN-IRELAND NONE

(A SUSPECTED IRISH REPUBLICAN ARMY GUNMAN KILLED A 50-YEAR-OLD
UNARMED SECURITY GUARD IN EAST BELFAST EARLY TODAY THE POLICE SAID)

~~~ text processing messages skipped   ~~~

>>> Beginning memory update ...

```
New features: EV6 (XX2) (S-ATTACK-PERSON)
ACTOR           ORG         IRA
                POL-POS     BAD-GUY
                AFFILIATION CATHOLIC
RESULTS         AU          HURT-PERSON
                HEALTH      −10
METHODS         AU          $SHOOT
LOCATION        AREA        WESTERN-EUROPE
                NATION      *NORTHERN-IRELAND*
```

~ Since these features are similar enough to the last story's features,
~ IPP concludes that they possibly will result in a new spec-MOP at
~ the end of the story.

Best existing S-MOP(s)—
S-ATTACK-PERSON—potential remindings: EV2 (XX1)

~~~ text processing messages skipped   ~~~

>>> Beginning final memory incorporation . . .

```
Feature analysis: EV8 (I-TERRORISM)
PLACE           NATIONALITY *NORTHERN-IRELAND*
                DIRECTION   EAST
                PLACE       BELFAST
ACTOR           ORG         IRA
                POL-POS     BAD-GUY
                AFFILIATION CATHOLIC
VICTIM          STATUS      UNARMED
                ROLE        AUTHORITY
                POL-POS     ESTAB
INSTANCES       AU          S-ATTACK-PERSON
LOCATION        AREA        WESTERN-EUROPE
                NATION      *NORTHERN-IRELAND*
```

~ Now the features which the 2 stories had in common are placed in 2 new
~ spec-MOPs. One spec-MOP is done for the I-TERRORISM MOP used in the
~ 2 stories, and the other is done for S-ATTACK-PERSON.

Creating more specific I-TERRORISM (SpM2) from events EV0 (XX1) EV8 (XX2)
with features:

```
ACTOR      (1)  ORG           IRA
                POL-POS       BAD-GUY
                AFFILIATION   CATHOLIC
VICTIM     (1)  ROLE          AUTHORITY
                POL-POS       ESTAB
INSTANCES (1)   AU            S-ATTACK-PERSON
LOCATION  (1)   AREA          WESTERN-EUROPE
                NATION        *NORTHERN-IRELAND*
```

Reminded of: EVO (XX1) (MOP creation)

Feature analysis: EV6 (S-ATTACK-PERSON)
```
PLACE           NATIONALITY   *NORTHERN-IRELAND*
                DIRECTION     EAST
                PLACE         BELFAST
I-MOP           I-MOP         I-TERRORISM
VICTIM          STATUS        UNARMED
                ROLE          AUTHORITY
                POL-POS       ESTAB
ACTOR           ORG           IRA
                POL-POS       BAD-GUY
                AFFILIATION   CATHOLIC
RESULTS         AU            HURT-PERSON
                HEALTH        − 10
METHODS         AU            $SHOOT
LOCATION        AREA          WESTERN-EUROPE
                NATION        *NORTHERN-IRELAND*
```

Creating more specific S-ATTACK-PERSON (SpM3) from events EV2 (XX1) EV6 (XX2) with features:
```
VICTIM     (1)  ROLE          AUTHORITY
                POL-POS       ESTAB
ACTOR      (1)  ORG           IRA
                POL-POS       BAD-GUY
                AFFILIATION   CATHOLIC
RESULTS    (1)  AU            HURT-PERSON
                HEALTH        − 10
LOCATION   (1)  AREA          WESTERN-EUROPE
                NATION        *NORTHERN-IRELAND*
I-MOP      (1)  I-MOP         I-TERRORISM
```

Reminded of: EV2 (XX1) (MOP creation)

>>> Memory incorporation complete

Story Representation:

```
** MAIN EVENT **
EV8 =
  MEM-NAME      I-TERRORISM
  ACTOR         GUNMAN OF THE IRA
  VICTIM        50 YEAR OLD IRISH UNARMED GUARD
  PLACE         EAST BELFAST
  INSTANCES
   EV6 =
    MEM-NAME    S-ATTACK-PERSON
```

```
      ACTOR            GUNMAN OF THE IRA
      VICTIM           50 YEAR OLD IRISH UNARMED GUARD
      RESULTS
       EV5 =
         MEM-NAME     CAUSE-DEATH
         ACTOR        GUNMAN OF THE IRA
         VICTIM       50 YEAR OLD IRISH UNARMED GUARD
         HEALTH       -10
         PLACE        EAST BELFAST
      METHODS
       EV7 =
         MEM-NAME     $SHOOT
         ACTOR        GUNMAN OF THE IRA
         VICTIM       50 YEAR OLD IRISH UNARMED GUARD
         PLACE        EAST BELFAST
    TIME        TODAY
```

4361 msec CPU (0 msec GC), 17000 msec clock, 6695 conses
NIL

*(parse xx3)

Story: XX3 (1 12 80) NORTHERN-IRELAND NONE

(A GUNMAN SHOT AND KILLED A PART-TIME POLICEMAN AT A SOCCER MATCH
SATURDAY AND ESCAPED THROUGH THE CROWD TO A WAITING GETAWAY CAR
COMMA POLICE SAID)
~~~ text processing messages skipped ~~~

>>> Beginning memory update . . .

New features: EV10 (XX3) (S-ATTACK-PERSON)
```
ACTOR         POL-POS       BAD-GUY
METHODS       AU            $SHOOT
LOCATION      AREA          WESTERN-EUROPE
              NATION        *NORTHERN-IRELAND*
```

~ Now that there is a spec-MOP in memory, and since the features extracted
~ from the new story agree with the features in the spec-MOP, it is chosen
~ instead of the default S-ATTACK-PERSON MOP to represent the story.
~ IPP predicts that the feature in the spec-MOP which are not in the story
~ so far will occur later on in the story.

Best existing S-MOP(s)—
SpM3—potential remindings: EV6 (XX2)

Predicted features (SpM3)
```
RESULTS       HEALTH        -10
              AU            HURT-PERSON
ACTOR         AFFILIATION   CATHOLIC
              ORG           IRA
VICTIM        POL-POS       ESTAB
              ROLE          AUTHORITY
I-MOP         I-MOP         I-TERRORISM
```
~~~ text processing messages skipped ~~~

>>> Beginning final memory incorporation . . .

Feature analysis: EV10 (S-ATTACK-PERSON)

SCENES	AU	SS-ESCAPE
VICTIM	OCCUPATION-TYPE	
		PART-TIME
	GENDER	MALE
	ROLE	AUTHORITY
	POL-POS	ESTAB
RESULTS	AU	HURT-PERSON
	HEALTH	− 10
ACTOR	POL-POS	BAD-GUY
METHODS	AU	$SHOOT
LOCATION	AREA	WESTERN-EUROPE
	NATION	*NORTHERN-IRELAND*

~ Again, the story is placed in long-term memory, represented as a pointer
~ to the best existing spec-MOP (in this case, SpM3) and a list of features
~ found in the story but not in the spec-MOP.

Indexing EV10 (XX3) as variant of SpM3

Reminded of EV6 (XX2) (Last MOP reference)

Reminded of EV6 (XX2) (varies from MOP in same way)

~ Since the spec-MOP had one feature which the story did not have
~ (ACTOR ORG IRA), this feature is inferred.

Adding default feature ACTOR ORG IRA to EV10

>>> Memory incorporation complete

Story Representation:

** MAIN EVENT **
EV10 =
  MEM-NAME   S-ATTACK-PERSON
  ACTOR   GUNMAN OF THE IRA
  VICTIM   PART-TIME POLICEMAN AT SOCCER MATCH
  METHODS
    EV9 =
    MEM-NAME $SHOOT
    ACTOR   GUNMAN OF THE IRA
    VICTIM   PART-TIME POLICEMAN AT SOCCER MATCH
  RESULTS
    EV11 =
    MEM-NAME   CAUSE-DEATH
    ACTOR   GUNMAN OF THE IRA
    VICTIM   PART-TIME POLICEMAN AT SOCCER MATCH
    HEALTH   − 10
  TIME   SATURDAY
  SCENES
    EV12 =
    MEM-NAME   SS-ESCAPE
    ACTOR   GUNMAN OF THE IRA

```
7260 msec CPU (2152 msec C), 28000 msec clock, 7252 conses
NIL
*
Interrupt (Help = ?):  ↑ X
@pop
[PH: Termination. 26-Jan-81 3:45PM. PS:<DEMO>IPP.LOG.1]
```

Again, IPP is meant to illustrate how a program can augment its knowledge through experience. This learning ability will be ever more important as people make greater demands on expert systems.

4. Adapting to Failures

What happens when a new problem arises? What happens when the program makes a mistake? How can the program adapt to failure?

Our work in modeling memory had suggested that failure is a critical part of learning and memory organization (Schank, 1981). When a person encounters an unexpected event, he tries to explain it. This explanation process is a basic learning mechanism.

There are numerous ways in which a plan can fail. Recognizing failures leads to a further refinement and respecification of the plan. Furthermore, if the failure can be explained or accounted for through another means, then the plan can be augmented as well.

IPP not only added new episodes to memory and formed generalizations, but it could also handle expectation failures. For example, some of IPP's generalizations were merely coincidences; two stories may have had similar details due to chance, rather than due to some intrinsic regularity in the world. For example, the program read two stories in which two people were killed in a Latin American country. IPP developed the rule that killings in that country would always come in pairs. When subsequent stories came in that did not adhere to this rule, IPP was forced to reconsider it and ultimately rejected it.

This adaptive behavior is another desirable attribute of an advisory system. When a new case comes in which does not quite fit with previous experience, the program must be able to change and adapt its knowledge-base.

5. Inferring the user's goals

Part of building a memory of cases entails building models of the clients themselves and their requirements. An advisory system must be helpful and convenient to use. The program should have the ability to infer the goals of the client with a minimum training with the client. The program should be able to handle a variety of client goals and to adapt its advice to the disparate needs of the users.

There are numerous types of goals that a client may have. Some goals overlap, and others conflict. Typical goals may include:

- maximize profits
- minimize overhead
- minimize risks
- reward productive employees
- fire nonproductive employees
- build market share
- maintain public trust
- don't break the law
- protect the safety of your employees
- minimize taxes
- don't divulge proprietary information

These are fairly general goals for business. An advisory system should have knowledge of these default goals and how they might interact. For a specific client, the program would need to be aware of additional specific goals as well as special ordering and relationships for the default goals. For example, a high-tech business might be much more concerned about proprietary information than a service business, which in turn might be more concerned with its public image.

We have done work in text understanding which demonstrated the need to model goals. The POLITICS program (Carbonell, 1979) simulated a conversation with a United States senator on a politically sensitive issue. POLITICS could adopt a political perspective in the conversation, either conservative or liberal, and evaluate the import of any news event on the basis of this viewpoint. The program used a series of goal trees which order political goals or beliefs into a hierarchy of relative importance.

The following is a sample output of the POLITICS program, with an input news story analysed from a conservative point of view.

```
*(INTERPRET US-CONSERVATIVE)

INPUT TEXT: Russia massed troops on the Czech border.

INTERPRETATION COMPLETED, READY TO ACCEPT QUESTIONS.

Q1:   Why did Russia do this?
A1:   BECAUSE RUSSIA THOUGHT THAT IT COULD TAKE POLITICAL CONTROL OF
      CZECHOSLOVAKIA BY SENDING TROOPS.

Q2:   What will Russia do next?
A2:   RUSSIA MAY TELL ITS TROOPS TO START THE INVASION.
```

Q3:   What else can Russia do?
A3:   RUSSIA MAY WAIT FOR CZECHOSLOVAKIA TO GIVE IN TO RUSSIAN
      DEMANDS.

Q4:   What happens if Czechoslovakia does not give in?
A4:   RUSSIA MAY DECIDE TO START THE INVASION.

Q5:   What should the United States do?
A5:   THE UNITED STATES SHOULD INTERVENE MILITARILY.

Q6:   What else can the United States do?
A6:   THE UNITED STATES CAN DO NOTHING, OR IT CAN INTERVENE
      DIPLOMATICALLY BY CONFERRING WITH RUSSIA ABOUT CZECHOSLOVAKIA.
      THESE ALTERNATIVES ARE BAD FOR THE UNITED STATES.

POLITICS can take the same news headline and interpret it quite differently from a liberal perspective. The main difference between these two modes of the program is the goal base.

This program comes fairly close to what one would like to see in an advisory system. POLITICS is an example of subjective understanding— understanding the meaning of an event from a particular perspective. It is this ability which is important for advisory systems. The memory-based expert system must be able to analyze a situation in terms of the goals and needs of the client.

6. Summary

We have discussed some problems in devising robust expert advice systems. Our model in this design is human expert performance. Following that psychological model, we are interested in developing expert programs with the following features:

- experience based memory to enable incremental learning
- failure driven memory to adapt to unexpected situations
- client goal modeling to provide advice tailored to the needs of the user

Current work at the Yale AI Project focuses on these problems. Our previous work in these areas in natural language processing suggests ways of addressing these problems. We feel that these problems will become more important with the need to develop high-performance expert systems.

7. References

Carbonell, J. *Subjective understanding: Computer models of belief systems.* Unpublished doctoral dissertation, Yale University, 1979.

Cullingford, R. *Script application: Computer understanding of newspaper stories* (Research Report #116). Unpublished doctoral dissertation, Yale University, 1978.

DeJong, G. *Skimming stories in real time: An experiment in integrated understanding.* Unpublished doctoral dissertation, Yale University, 1979.

Lebowitz, M. *Generalization and memory in an integrated understanding system.* Unpublished doctoral dissertation, Yale University, 1980.

Samuel, A. L. Some studies in machine learning using the game of checkers. In E. A. Feigenbaum, & J. Feldman (Eds.), *Computers and thought.* New York: McGraw-Hill, 1963. [Also in IBM Journal of Research and Development (1959).]

Schank, R. C. *Fundamental studies in computer science. Volume 3: Conceptual Information Processing.* North-Holland, Amsterdam: 1975.

Schank, R. C. Failure-driven memory. *Cognition and Brain Theory,* 1981, *4*(1), 41–60.

Schank, R. C. *Dynamic memory: A theory of learning in computers and people.* New York: Cambridge University Press, 1982.

Schank, R. C., Lebowitz, M., & Birnbaum, L. *Integrated partial parsing* (Tech. Rep. 143). Department of Computer Science, Yale University, 1978.

Shortliffe, E. H. *Computer-based medical consultations: MYCIN.* New York: American Elsevier, 1976.

Wilensky, R. *Understanding goal-based stories* (Research Report #140). Unpublished doctoral dissertation, Yale University, 1978.

FROM

ARTIFICIAL INTELLIGENCE
APPLICATIONS FOR BUSINESS

PROCEEDINGS OF THE
NYU SYMPOSIUM,
MAY 1983

EDITED BY
WALTER REITMAN
BBN LABORATORIES

ABLEX PUBLISHING CORDORATION
NORWOOD, N.J. 07648

Market Trends in Artificial Intelligence

Howard Austin
KNOWLEDGE ANALYSIS, INC.

This report describes trends now emerging in the AI marketplace. The major categories of market participants are illustrated by case studies, with special attention given to the reasons which motivated them to join the AI movement and the results they have achieved to date. The report concludes with an analysis of the market trends observed and speculations about likely future scenarios.

1. Introduction

This chapter gives an overview of current activity in the AI marketplace. It is addressed to a lay audience which is assumed to be business-oriented. The presentation uses representative case studies to illustrate the major categories of market players. Summaries of related activity are given for each category and a crude estimate of overall market size is offered. The paper concludes with an analysis of trends and problems which are now emerging in the AI marketplace and a set of predictions about likely future scenarios.

The cases used in this presentation are well known within the AI community but the analysis offered is original. The mistakes, therefore, are solely the author's. The reader will note that the presentation emphasizes three major perspectives on the current marketplace: (a) The author's personal experience as an AI employee, consultant and startup participant, (b) the Wall Street perspective, and (c) the perspective of a concept called The Repetition Spectrum which is offered as a way of thinking about how new inventions are propagated through a society and how businesses position themselves to take advantage of that spread. Topics not discussed in this overview include activities such as Robotics, Vision, Pattern Recognition, Computer Networks, etc., which are generally regarded as "AI spinoffs".

2. The Repetition Spectrum

The concept of a repetition spectrum is as follows:

> In any society, all fundamentally new ideas are generated by a handful of creative individuals. These ideas are taken up by those who come in contact with the original sources and are "repeated" by elaboration and incorporation into the agendas of the new "owners". If the idea has sufficient persuasive power, its essential core is preserved and spread, repetition by repetition, in larger and larger circles until it completely saturates the society. Individuals and institutions tend to position themselves in characteristic positions along the resulting spectrum. The essential question to ask then, when assessing new technologies, is: where are we on the repetition spectrum?

Figures 1 and 2 illustrate this point for the currently popular technology of expert systems.

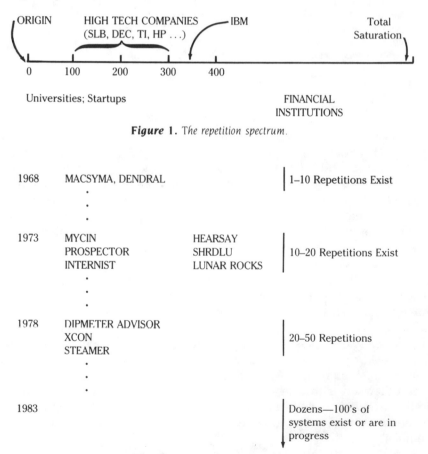

Figure 1. *The repetition spectrum.*

1968	MACSYMA, DENDRAL		1–10 Repetitions Exist
1973	MYCIN PROSPECTOR INTERNIST	HEARSAY SHRDLU LUNAR ROCKS	10–20 Repetitions Exist
1978	DIPMETER ADVISOR XCON STEAMER		20–50 Repetitions
1983			Dozens—100's of systems exist or are in progress

Figure 2. *Expert system example.*

Another key insight from this model is that businessmen and intellectuals tend to be on opposite ends of the repetition spectrum. An intellectual is a person who wants to do a million different things once, because that's the best way to understand the world. A businessman is a person who wants to do one good thing a million times, because that's the best way to make money. Intellectual excitement occurs mostly on the left end of the spectrum, financial leverage mostly on the right. This observation leads to useful questions about the motivation of AI businessmen, the degree of focus found in their businesses and the packaging of their products. We will have occasion to return to these questions in the case studies which follow.

3. Fortune 500 Activity (The Schlumberger Case)

The first category to be considered consists of Fortune 500 corporations who have started major AI groups of their own. Xerox was first in this regard (early 1970s) but was so far ahead of the competition that the effort was essentially university-like in character. Schlumberger, with its highly publicized 1978 entry into AI, is usually credited with initiating the current wave of interest and is offered here as representative of Fortune 500 activity. The following discussion summarizes the motives, activities and results of the Schlumberger AI effort.

3.1 Schlumberger Motives

Schlumberger is an oil service company. Its basic business is collecting data from oil wells drilled by other contractors. At the scale Schlumberger deals with, a rock internally is like a sponge. Externally, under the massive forces of geology, rocks behave like taffy. The oil flows through the pores of the sponge (which is typically sandstone or limestone). Significant accumulations of oil occur only where geological formations interact to block further migration from above, below and on all sides, thereby forming an "oil trap." (See Figure 3.)

Schlumberger's goal is to help its clients locate one of the 50 or so kinds of known trapping situations and evaluate the hydrocarbon potential. To do so, Schlumberger has developed a family of electrical, acoustical and nuclear probes which are lowered into wells by wireline cable. Each probe measures some petrophysical parameter (e.g., resistivity) which is useful in constructing a picture of the subsurface geology and its hydrocarbon content. The measurements are recorded on long pieces of paper known as well logs, and are interpreted by human experts skilled at inferring geological information from the masses of squiggly lines. Therein lies the motive for starting an AI group. Schlumberger's ability to generate "interesting" new measurements currently exceeds, and threatens to exceed by

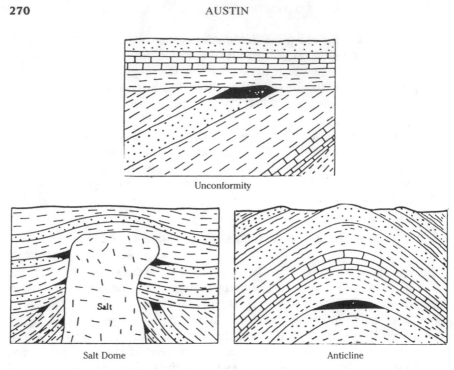

Figure 3. *Commonly occurring oil traps.*

orders of magnitude, the ability of the human interpreters to process those measurements. Artificial Intelligence, with its promise of "intelligent" mechanical interpreters, offers a major new approach to a problem that Schlumberger absolutely has to solve, in order to keep growing, namely, that *lack of interpretation limits sales.*

3.2. *Schlumberger* **Activities**

To address this problem, Schlumberger began in 1978 to recruit and equip an AI group. The problem this group chose to work on was that of creating an expert system for dipmeter interpretation. The dipmeter is a tool used to infer the tilt (i.e., dip) of beds encountered in the subsurface. The dipmeter tool records four resistivity curves. These curves are generated by pressing four current producing electrodes against the inner face of the borehold as the tool is withdrawn from the well. Correlations of the relative shifts of "the same resistivity pattern", coupled with information about tool orientation, allow the interpreter to reconstruct the depth, dip, and orientation of the beds encountered by the tool. In actual practice the curve correlation process is performed by a machine, producing an intermediate stage representation called a "tadpole plot" which is then interpreted by a human expert. (See Figure 4.)

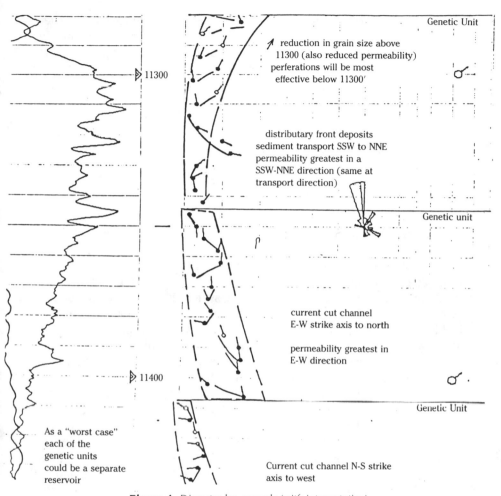

Figure 4. *Dipmeter log example (with interpretation).*

This latter stage, the interpretation of "tadpole plots" has been regarded by many observers as quite unscientific, particularly since the observers could seldom identify the process by which a conclusion was reached and the recognized experts frequently disagreed among themselves. Nevertheless, the problem seemed well suited to AI techniques so work began on the knowledge extraction problem. It quickly became apparent that the expert under study used knowledge that did not readily lend itself to representation in the standard IF-THEN rule format and, moreover, operated at many different levels in a given interpretation. Worse yet, the expert seemed to guess the answer and then impose it on the data to check the degree of fit. This approach defies the basic assumption of many pattern-directed inference systems in which interpretation proceeds from signal to conclusion, rather than vice versa.

On the other hand, it was also clear that the expert *did* have a consistent (though previously private) overall system and that this system did involve the use of a number, around 50, of IF-THEN rules of the type used by previous expert system efforts. These observations led to a decision to proceed with the construction of a prototype expert system of the standard kind with the clear understanding that the resulting model would be a shallow, first pass theory of a previously mysterious skill. It was felt that such a system would still have merit as a demonstration of Schlumberger's competence in AI and as a basis for future work. Accordingly, an expert system— The Dipmeter Advisor, was designed, implemented, and successfully tested, complete with approval by the dipmeter expert and by Schlumberger management.

3.3. *Schlumberger* Results

Later versions of the Dipmeter Advisor were given expanded rule sets and re-implemented on a variety of new computers, including personal LISP machines (e.g., DOLPHINS). The resulting system is now in field evaluation in preparation for introduction to Schlumberger's worldwide network of Field Log Interpretation Centers (F.L.I.C.). In the research center, efforts are now being made to address the more fundamental issues bypassed in the first implementations, including an overhaul of the basic representation scheme. Related efforts are applying AI techniques to more broadly based systems capable of utilizing data from a great many sources, instead of just one tool. Efforts are also underway to create AI-based programming and log interpretation environments which will improve the laborious processes by which software is written and new log interpretation models are created. These efforts have been successful enough to generate rumors that Schlumberger has ordered large numbers of Lisp machines and will eventually install one for every log interpreter. Consideration has also been given to replacing Schlumberger's thousands of truckborne PDP-11 computers with a Lisp machine configuration designed to deliver better log interpretation at the well site.

In addition to the efforts of the Ridgefield, Connecticut, research center, Schlumberger has established a second major AI group in its newly acquired Fairchild subsidiary and has acquired access to yet another group by its purchase of a 15% share of Bolt, Beranek & Newman. These extensive commitments, and the style in which they were done, have established Schlumberger as a world leader in commercial AI activity.

3.4. *Schlumberger's* Cost Estimates

The cost of Schlumberger's commitment is not publically available but can be estimated as follows:

RIDGEFIELD	1) 10 professionals @ 125K/yr. × 5 yrs. =	$ 6,250,000
	2) Dec 2020 + 15 DOLPHINS =	1,000,000
FAIRCHILD	1) 15 professionals @ 125K/yr. × 3 yrs. =	5,625,000
	2) AI computing facilities =	2,000,000
BBN	1) 300,000 shares @ $22/share =	6,600,000
		$21,475,000

Despite the crudeness of these estimates, the cost is still clearly in the 10–20 million range.

The potential return on this investment is equally enormous but is not easy to calculate, since it is mostly invisible to the outside world. Schlumberger chose to focus its AI effort on internal projects which improve its competitiveness in its basic business. Some slightly out-of-date statistics serve to illustrate the possibilities. In 1978, the dipmeter log accounted for 9% of wireline sales, an amount which totaled 200 million dollars for that one tool. The dipmeter, at that time, was sold on only 1 well out of 4, due primarily to the lack of qualified dipmeter interpreters. If the Dipmeter Advisor System can extend the dipmeter sales ratio to 2 wells out of 4, that would generate an additional 200 million per year in revenues for this one tool. Similar arguments can be made for other tools and for the benefits which arise from having demonstrated a general solution to the critical growth limit posed by the interpretation bottleneck. By using AI to become even more competitive, Schlumberger guarantees the continuation of its extremely lucrative basic business.

Table 1
Summary of Fortune 500 AI Activity

Fortune 500 Company	Application Area
Xerox	Office of the Future
Schlumberger	Well Log Interpretation
Texas Instruments	Seismic, Education, Robotics, VLSI
DEC	VAX Configuration, Fault Diagnosis
IBM	Fault Diagnosis, Database Interface
Hewlett-Packard	Instrument Interpretation, VLSI
Atari	Intelligent Game Design
General Motors	Robotics, Vision
Bell Labs	Natural Language, Database
Westinghouse	
General Electric	
Lockheed	
Amoco	Oil Field Exploration Service
Boeing	
Fairchild (Schlumberger)	VLSI, Vision
Tektronics	
Standard Oil of Ohio	Oil field, chemical analysis

4. Wall Street Activity (The F. Eberstadt Case)

The next major category of AI market participant to be examined is that of
the Wall Street investor. The players in this category are most readily understood in terms of their relationship to the stock market, that being the
central metaphor which dominates their lives. The F. Eberstadt case study
illustrates several aspects of this metaphor as it relates to the AI industry.

4.1. F. Eberstadt Background and Motives

The first question, of course, is what is F. Eberstadt? The simple answer
is that F. Eberstadt is an investment bank cum research boutique which
wants to be the investment banker for AI. The more detailed answer is that
banks come in flavors. Retail banks loan money to individuals. Commercial
banks loan money to corporations. Investment banks don't make loans at
all, but instead, provide services which help existing businesses raise capital
by selling stock in the public or private markets. The emphasis, for investment bankers, is on the phrase "existing businesses", the province of
"would-be business" being reserved for venture capitalists.[1]

F. Eberstadt's position in this galaxy is that of a research intensive
investment bank which focuses on selected areas of high technology. Within
this focus it further specializes in the private placement market (but offers
the full range of investment services as required). The private placement
activity consists of matching the investment needs of large private investment pools (e.g., pension funds) with the capital needs of carefully
screened technology startups. These companies, in Eberstadt's judgement,
are the cream of the post-venture capital crop and could easily go public
if they chose to do so. For a variety of reasons, (including a desire for
greater freedom), they choose to remain private, but still need to raise
capital to continue growing, thus creating one side of the market need
which Eberstadt fills.

The managers of major money pools provide the other side of the
market need. They want to find tomorrow's hot stock today. However, being
specialists in money management they don't have time to be specialists
in technology and so they rely on institutions like F. Eberstadt to alert them
to the implications of APPLE or GENENTECH and assist them with the preliminary screening of investment candidates.

In order to reliably make such judgements, F. Eberstadt maintains a
large, in-house, research staff which follows selected industries (e.g., medical
technology, chemicals, energy). These analysts serve as sounding boards

---

[1] In reality most major money center banks offer all of these services under one roof. Functionally and historically, these divisions are essentially correct, but are now being further
eroded by "financial supermarket" entities like Merrill Lynch and Sears Roebuck.

for and prime movers of the investment banking activities. In addition, their research reports are marketed separately, providing a major source of revenue. This revenue comes from institutional clients who pay for the research services in the traditional way, by doing their brokerage business through F. Eberstadt's institutional brokerage arm.

4.2. F. Eberstadt Activities

F. Eberstadt is a knowledge broker. Its reputation for being able to spot winners is its stock in trade. That ability in turn, is based on the expertise of its in-house research staff and their ability to understand the investment implications of new technologies, like medical instrumentation, biotechnology and artificial intelligence, to name three areas of recent interest. Their general strategy for approaching new areas is to cover the waterfront, i.e., understand the basic ideas in the field, get to know the academic experts, attend conferences, visit the existing companies, talk with the people who financed them, and so on. Then, having educated themselves, they wait for "their pitch" to emerge. The baseball analogy is very relevant here. The stock market game is so complicated that no one can afford to swing at every pitch. The more successful players tend to have a small number of ideal investment situations, for which they wait very patiently, on the grounds that they will do very well when one of those situations comes along.

F. Eberstadt's notion of an ideal investment, crudely sketched, looks something like this.

1. The company's products give it a position of continuing dominance in a new and important field.
2. There is a realistic potential for explosive growth (minimum goal = $100 million annual sales).
3. Sales are currently in the 5–10 million range.
4. Management and staff are top quality and preferably experienced.
5. The valuation placed on the company is conservative.
6. The company was previously financed by top flight venture capitalists (e.g., Adler, Rock, et al.).
7. The company is willing to consider the private placement strategy.
8. The overall situation "feels right".

These criteria are obviously heuristic in nature; no one invests solely on the basis of a checklist. Nevertheless, guidelines are useful. It is interesting to note, for example, that none of the widely known AI companies quality on item 3 (sales of 5–10 million). Indeed, all existing AI companies are small potatoes in the world of big time finance, a fact that is frequently bruising to some of the massive egos found in the field.

There *are* companies which satisfy Eberstadt's criteria, however, the most notable example being a company called DIASONICS, which is the industry leader in Nuclear Magnetic Reasonance (NMR) imaging devices. Eberstadt did several rounds of private placement financing for DIASONICS, including the biggest private placement in history, and was a lead investor in the syndicate which just took DIASONICS public. DIASONICS at the time had annual sales of 138 million and was given in the offering an overall valuation in excess of one billion dollars.

4.3. Eberstadt Results

What does F. Eberstadt have to show for its AI activities to date? Predictably, their initial contact with AI came from one of the research analysts, a person named Phil Meyer. Meyer is widely recognized, on Wall Street and in the oil industry, as the world's leading Schlumberger analyst. He showed up at an obscure talk about a thing called a DIPMETER ADVISOR and started asking a lot of questions. One thing led to another, which led to introductions, which led, largely on Meyer's prompting, to a decision by F. Eberstadt to do a full-scale investigation of AI. This investigation proceeded, in the fashion previously outlined, culminating in a recent MIT conference on AI, cosponsored by F. Eberstadt and attended by over 700 participants.

Meyer, in the meanwhile, had stolen the march on his competitors, by being the first industry analyst to understand and issue a research report on the strategic significant of Schlumberger's AI commitment. Meyer's interest in, and enthusiasm for, AI also contributed to the formation of RES-TECH, an oil service startup which hopes to ultimately use AI technology in the reservoir/log interpretation business.

On the investment banking side, discussions with members of the AI community had the side effect of providing an intellectual framework for evaluating related deals (e.g., MICROPRO/personal computer software). The most significant direct result, to date, has been F. Eberstadt's participation in a private placement for a company called DAISY, which makes "AI-like" workstations which dramatically improve the circuit design process. CAD/CAM companies were the darlings of Wall Street in the 1970s. Companies like DAISY promise equally exciting reults for the 1980s. DAISY, interestingly enough, uses AI technology to create "smart" software/circuit design assistants but refuses to publicly label it AI. The implications of DAISY's attitude toward and use of AI are interesting enough to make it the subject of the following section on AI startups. The conclusion here, and for F. Eberstadt, is that although it may be a little early (from the Wall Street perspective) for "pure" AI plays, the initial promise of AI is rapidly bearing fruit in the form of applications like DAISY, which have here and now validity. Table 2 summarizes financial institutions which are known to have made AI investments.

Table 2
Summary of Known AI Investors*

Company	AI Investments	Location	Contact
1. Alan Patricoff Associates	Symbolics	New York, NY	John Baker (212) 753–6300
2. American Research and Development	C*T, Symbolics LOGOS	Boston, MA	Gene Pettinelli (617) 523–6411
3. Bank of Boston High Technology Lending Group	Symbolics C*T AI, CORP, LOGOS	Boston, MA	Tom Farb (617) 434–5570
4. Bessemer Venture Partners	LOGOS	New York, NY	William Buescher (212) 708–9100
5. F. Eberstadt & Co.	DAISY	New York, NY	Bill Janeway (212) 480–1359
7. Fairfield Venture Management	C*T, Symbolics	Stamford, CT	Randy Lunn (203) 358–0255
8. First Venture Corp. of Oklahoma	LOGOS	Bartlesville, OK	J.R. Tinkle (918) 333–8820
9. Adler & Company	DAISY	New York, NY	Steve Clerman (212) 986–3010
10. G.E. Venture Capital Corp.	Symbolics, C*T	Fairfield, CT	Jim Fitzpatrick (203) 357–4100
11. General Instrument	Symbolics	New York, NY	Lewis Solomon (212) 765–9373
12. Industrial Dev. Partners	LOGOS	Germany	Dr. Peter Hengel
13. Lambda Fund (Drexel Burnham)	LOGOS	New York, NY	Anthony Lamport (212) 480–6000
14. Memorial Drive Trust	Symbolics C*T	Cambridge, MA	Jean de Valpine (617) 864–5770
15. Northwest Growth Fund	C*T		
16. Oak Management Corporation	AI Corp.	Westport, CT	Ginger More (203) 226–8346
17. Sevin Rosen Partners	C*T	Richardson, TX	L.J. Sevin (214) 960–1744
18. TA Associates	AI Corp.	Boston, MA	Jackie Morby (617) 725–2300
19. Vista Venture Corp.	AI Corp.	Stamford, CT	Jerry Bey (203) 359–3500
20. Aeneus (Harvard Mgt. Corp.)	C*T	Boston, MA	W. Haberkorn (617) 423–4250

*As of May 1983.

5. Startup Activity

The next category of market participant to be considered is that of AI startup activity. The first problem here is in distinguishing between "purely" AI startups and related endeavors (e.g., Three Rivers, LOGO Systems). The test used here is that either the ideas or the principals involved must be closely associated with the AI community. The second problem is in picking a representative example, because there aren't any good ones yet, at least from the Wall Street perspective (meaning that none of the existing 20 or so companies are obvious candidates for the $100 million/year sales level). Consequently, the discussion will focus here on DAISY Systems, a choice that Wall Street would regard as a good AI play but one that may come as a surprise to members of the AI community. The reasons, in both cases, are instructive.

5.1. Don't Call It AI

DAISY Systems is in the computer-assisted engineering business (C.A.E.). Its chief product is the LOGICAN workstation. In conventional engineering, CAD systems help turn circuit diagrams into chip layouts, and CAM systems help turn the resulting chips into finished products. In the envisioned new design era, the LOGICAN system will serve as an "intelligent assistant" which helps the engineer convert design ideas into circuit diagrams, thereby achieving another major step towards a totally automated design process.

There are several interesting aspects to DAISY's overall approach. The first is an oft repeated story told by DAISY's founder, Aryeh Feingold. Feingold, and many key members of the DAISY staff, are former employees of Intel, where they were responsible for the design and programming of some of Intel's most powerful microprocessors. As such, they are obviously first rate hardware and software designers. Frequently, according to Aryeh, these wizards would say "Look, here's a place where we can use AI" (in the LOGICAN). Feingold's response would always be to slap their hands and say "Stop. *Don't call it AI.* Call it advanced programming". He goes on to explain that he didn't set out to start an AI company, or any other kind, for that matter. He set out to solve a problem and that problem happens to demand AI techniques (the search space is much larger in C.A.E. than in CAD/CAM). Furthermore, AI, in some circles, became an excuse for not getting things done and he wanted to avoid any hint of that. So he used the ideas, thereby adding his contribution to the repetition spectrum for expert systems, but he changed the name.

Another important aspect of LOGICAN is that its efficacy is rumored to be such that serious designers will have to have one (or something equivalent) in order to remain competitive. If this is substantially true (allowing for hype) that means the market is enormous, potentially consisting of every design engineer. Time will tell on that score. However, it is already clear that the LOGICAN addresses an important, hard problem (designing faster), which gives enormous leverage if successfully solved, and which is of interest to a large number of people.

DAISY's packaging is also instructive. Although the LOGICAN is at least functionally equivalent to an expert system, the software is not sold separately, but instead, is delivered as an integral part of a DAISY-designed professional workstation (compatible, of course, with existing CAD/CAM systems). There's nothing special about the workstation, except that it's cheap for DAISY to build it. The net effect of this "bundled" packaging is that DAISY can behave like a computer manufacturer, instead of like a software house. The ramifications of this change affect pricing and marketing strategies, in fundamental ways. It is a transformation that AI vendors of expert system would do well to learn from. Recall, for example, IBM's success with a similar strategy before it was forced to unbundle.

The final point of interest about DAISY is that it is extremely well financed and well connected on Wall Street. The lead venture capitalist is Fred Adler, who is also chairman of the board. Adler has been involved in over 100 startups (including success stories such as Data General). He personally oversees a 150 million dollar venture capital pool. F. Eberstadt, which manages a private placement pool of equal size, has just completed an initial private placement for DAISY. The company has a unique product, a strong marketing orientation, a talented staff, no debt, a lead on the competition and sales are booming ($10 million in 1983). Sales are projected to reach 100 million within 5 years and, unlike many projections, will probably be achieved with room to spare. Table 3 gives a summary of related AI startup activity.

6. Market Size Estimate

The preceding case studies, with their peculiar mixtures of visible and invisible activities, should amply illustrate the difficulties involved in estimating the overall size of the AI market. The policy followed here is to identify the extremes of the range of plausible estimates. Table 3, for example, summarizes known AI revenues for 1982 and gives projected revenues for 1983.

The 1982 actual AI revenues, a sum equal to $18 million, provide the most conservative estimate of AI market size. This total does not include, for example, the revenues spent internally on AI projects by members of the Fortune 500. There are approximately 20 such companies, each of which spends, on the average, 2–3 million per year on AI, which leads to an es-

Table 3
AI Market Size Estimate*

Name	Location	Date Started	Capital Raised	Product	1982 Sales	1983 Sales	No. of Employees	Contact
1. Symbolics Inc.	Cambridge, MA	1980	14M	Lisp machines	3.5M	11.5M	216	Russ Noftsker (617) 864–4660
2. Three Rivers Inc.	Pittsburgh, PA	1974	N/A	PERQ Workstation	N/A	N/A	N/A	Edward Fredkin (412) 621–6250
3. LMI	Cambridge, Los Angeles	1980	1M	Lisp machines	1M	2M	20	Richard Greenblatt (617) 876–6819
4. Teknowledge	Palo Alto, CA	1981	1.5M(est.)	Expert systems	1.2M	3–5M	50	Lee Hecht (415) 327–6600
5. Applied Expert Systems	Cambridge, MA	1982	Sub. of INDEX	Expert systems	N/A	N/A	10 (est.)	Fred Luconi (617) 492–7322
6. Intelligenetics	Palo Alto, CA	1980	2M	K.E. tools	N/A	N/A	26	Tom Kehler (415) 328–4870
7. Smart Systems Technology	Alexandria, VA	1980	Self-capitalized	AI tools, seminars	300K	750–900K	5	Eamon Barret (703) 448–8562
8. Brattle St. Research	Boston, MA	1981	N/A	Expert systems	N/A	N/A	3	John Clippinger (617) 720–0051
9. DAISY Systems	Sunnyvale, CA	1980	8.8M	CAE Workstation	4.6M	10M	136	Aryeh Feingold (408) 773–9111

Company	Location	Year		Products			Employees	Contact
10. AI Corp.	Waltham, MA	1975	N/A	Nat. Lang. Systems	2M	4M	45	Larry Harris (617) 890–8400
11. Cognitive Systems	New Haven, CT	1979	2M	Nat. Lang. Systems	250K	1M	20	Roger Schank (203) 773–0726
12. Symantec	Sunnyvale, CA	1981	N/A	Personal Computer Software	(R & D Phase)		14	Gary Hendrix (408) 737–7949
13. Machine Intelligence Corp.	Sunnyvale, CA	1979	0.7M (est.)	Robotics, Nat. Language	N/A	N/A	90	Charles Rosen (408) 737–7960
14. Computer * Thought	Plano, TX	1981	N/A	Training systems	N/A	N/A	20(est.)	Mark Miller (214) 424–3511
15. LOGO Systems Inc.	Cambridge, MA Montreal, Canada	1981	N/A	Personal Computer Software	5M (est.)	10M(est.)	50	Vitold Jordan (514) 631–7081
16. LOGOS Computer Systems	Bedford, MA	1969	N/A	Lang. translation	200K	500K–1M	75	Charles Staples (617) 275–4334
17. Language Technology Co.	Boston, MA	1982	N/A	Cobol Regimenter	N/A	N/A	5	Erik Bush (617) 367–9685
18. Xerox	Los Angeles, CA	—	—	Lisp machines	N/A	N/A	—	Bo Bomeisler (213) 351–2351
TOTALS			30M		18M	42.7M	785	

*As of May 1983.

timate of 40–60 additional millions of AI dollars. The conservative total also does not include the estimated 30–40 million invested in AI by venture capitalists and other finances, or the university AI budgets (some of which exceed 5 million/year).

At the other extreme of estimation, the Japanese government plans to spend 45 million over the next three years on their 5th generation (AI) computer project. A consortium of 8 private companies will spend an additional 400 million during the same interval with 800 million more targeted for the rest of the decade. The U.K. just announced the funding of a 200 million pound government sponsored AI project with an additional 150 million pound contribution by private industry. Similar projects, of unknown magnitude, are now underway in Italy, France, and West Germany.

In the U.S., ARPA has, for some time, been spending 20 million annually on AI. It has just received an additional 20 million targeted for AI and is seeking 50 million more for super-computer development, some of which will go to AI projects.

In summary, estimates of the size of the AI market range from $18 million actual revenues in 1982 to projections of well over one billion cumulative total for the next 5 years.

7. Emerging Trends

Three kinds of trends are now visible in the AI marketplace: product trends, application trends and side effect trends.

7.1. Trend to Tool Based Strategies

The most important product trend is a pronounced, apparently industry wide shift, towards a tool based strategy for selling and doing AI. Schlumberger, for example, started out to make expert systems but finds itself contemplating a switch to using Lisp machines in all of its computing facilities. Expert system vendors like Teknowledge are finding it difficult to make large profits selling expert systems and are moving towards a tool based strategy. Competition between Lisp machine vendors such as Xerox, Symbolics, and LMI has heated up and threatens to get even hotter with the rumored entry of Texas Instruments and Hewlett-Packard, not to mention the Japanese. From the perspective of the repetition spectrum, the renewed emphasis on AI tools is both predictable and instructive. It's predictable because AI tools (such as Lisp) have been through many more repetitions than most other aspects of the technology and are therefore much better product candidates. It's instructive because the spectrum argument points out that expert systems, as currently packaged, tend to be one of a kind items and therefore should be either extremely expensive or unprofitable. Current results seem to confirm this view.

7.2. Internal Application Trend

Expert systems continue to be extremely popular but do not lend themselves to sale as external products. The trend is towards internal applications aimed at leveraging critical skills and technologies out to wider, but still internal, audiences. As currently constituted, expert systems are cost avoidance or performance enhancement systems rather than direct revenue generators. The most successful expert system to date, (DEC's XCON) has demonstrated savings of 10 million per year but is not intended for sale as an external product. One obvious strategy here is to charge on a "per use" basis. Some of the financial systems rumored to be under development will probably follow this route. Another approach is to repackage à la Daisy.

7.3. Interpretation Added Trend

Another theme visible in the marketplace is the use of AI technology to add an interpretation facility to an already existing data collection facility. Schlumberger is an example of this, as is the seismic interpretation effort at Texas Instruments. It is, undoubtedly, one of the chief motives behind the recent Hewlett Packard move into AI. A less well known example is the use of Helena Labs (Beaumont, Texas) of a chip-based expert system for the medical application of densitometer interpretation. This is an obvious growth area, as is the use of AI in intelligence related data interpretation (e.g., NSA, CIA).

7.4. Side Effects Trend

A counter argument to the charge that expert systems are inherently low repetition devices, is that successful expert system efforts always spawn a family of closely related systems. This is true of the Dipmeter Advisor at Schlumberger, which led to the more broadly based geological systems, and is also true of XCON at DEC which led to a family of sales support tools. This "seed" effect is reliable and should be incorporated into planning and marketing strategies by those who are buying and selling AI systems.

Two other important and commonly observed expert system side effects are the legitimization of previously private, heuristic knowledge and the educational benefits which arise from learning to see expertise as discrete, easily assimilable, chunks of knowledge. These side effects are frequently important enough to justify expert system projects on their own, independent of the benefits obtained from actually constructing the system. Obvious benefits of these side effects include the reduction of expert vs. expert disagreements, the codification of "accepted" knowledge about important domains (e.g., medicine) and the possibility of dramatically improved learning times.

7.5. Emerging Problems

The problems now emerging in the market place have been extensively discussed by others and so are merely summarized here. The chief problem, of course, is the shortage of AI talent. This shortage means that would be participants must either raid someone else or grow their own. Raiding is preferred by a wide margin. The major AI universities are now demoralized because of the raiding, their relatively low pay scales, and the lack of appropriately renumerative links between the ideas generated by their researchers and the profits generated by corporations using these ideas (corporations have exceedingly short memories in that regard). In terms of communication, there is still a great deal of hype, the AI systems building philosophy has only recently been well documented, and AI systems are *not* user-friendly, being mostly designed for use by the super-wizards found in AI centers.

In the expert systems area, knowledge extraction remains a central bottleneck, as do false expectations about system performance. (The systems are not as "smart" as their designers would have you believe.) The financial community, in particular, is prone to overestimate the importance of AI technology, or else to ignore it. This is partly due to their lack of experience with computing in general and partly due to their lack of good analogies for what AI is like. The most commonly quoted analogy (AI is like biotechnology) seems far off the mark. Businesswise, AI is much more like general purpose computing—round two, than like biotechnology. Better analogies are needed.

8. Scenarios of the Future

This paper concludes with a projection about the future of the AI marketplace. However, instead of concentrating on Japan's much discussed fifth generation computer project, or on obvious trends like the intelligence and military applications of AI, this projection focus on a less well appreciated aspect of the AI revolution, that of the "fun" industry.

In the future, there will be
only one industry: entertainment.
Marvin Minsky

This quote, like many of Minsky's, is outrageous but has more than a little truth in it. The man who seems to be making it come true is Nolan Bushnell. The world laughed at Bushnell when he founded Atari and starting selling PONG to the world in 1972. In 1976, Bushnell sold Atari (annual sales $26 million), pocketed 14 million and laughed all the way to the bank. Atari, with Alan Kay as its chief scientist, is now the largest customer for Symbolics Lisp machines. More intelligent games (e.g., learning) figure prominently in Atari's future plans.

The world laughed again when Bushnell announced he was going to use "Chucky Cheese Robot" and arcade games to sell pizza. Pizza Time Theatres now ranks third (2 spots ahead of Apple) on INC. Magazine's list of fastest growing companies. Annual sales exceed 100 million; there are now over 1000 franchises and a new one opens every four days (readers will recall that Radio Shack is the largest franchise chain with over 7000 stores; MacDonald's has around 4000).

Characteristically, Bushnell, still laughing, has moved on to a new enterprise, Catalyst Technologies. The plan here is to create something of an entrepreneurial training institute. Entrepreneurs will bring their inventions to Catalyst, which supplies the housing, capital generation, and management expertise typically needed by such enterprises. Successful projects, like the Androbot personal robot company, will have their products marketed through the Pizza Time Theatre distribution system. Bushnell's no-compete agreement with Atari is also about to expire so he's now free to re-enter the video game business. His objective in all of this?—*provide an alternative to the public school system.* As Bushnell says, "the only (intelligent) business to be in is the fun business". Laughs anyone?

The Artificial Intelligence Tool Box

Beau Sheil
XEROX SPECIAL INFORMATION SYSTEMS
PALO ALTO, CA 94304

Artificial intelligence is often claimed to be applicable to a variety of commercial applications. Here, we argue that a significant part of that applicability derives, not from the AI functionality per se, but from the underlying software technology. This software technology is examined, its genesis in the nature of AI programming is described, and the ways in which it differs from conventional software engineering are explained. The effectiveness of the AI programming technology for conventional application development first became clear when it was used to construct the information management environments in which the first AI applications were packaged. Building such environments is known to be extremely difficult using conventional software technology. But, in many ways, AI programming technology turns out to be ideally suited to this kind of development. The ability to build highly customized information management systems is itself an application opportunity of enormous leverage. Thus, in the short run at least, the application development power of AI's software technology may turn out to be just as important as the AI techniques for which it is nominally just a carrier.

Artificial intelligence (AI) is sometimes described as "the tool of the future". Describing something as a "tool" is an interesting rhetorical device. To begin with, it's very hard to falsify. Unless something is positively harmful (a judgement that some *have* passed on AI), it is difficult to argue that it couldn't be a tool (albeit perhaps for some very obscure situation that actually hasn't arisen yet). Further, if something is simply a tool (much less a "tool of the future") it doesn't have to have very many attractive or useful properties in the here and now. ("You wouldn't actually use this, but it's a tool that will be helpful in the development of something else that you might use later.") All kinds of deficiencies can be excused in a "tool". Yet, oddly enough, this unpromising platitude turns out to be the key to some surprising observations about the future application of AI technology. To begin with, it focuses attention on the *enabling* aspects of

AI technology, rather than on current capability, which is often limited. It raises the issue of what parts of AI might be useful as tools, what these tools might be useful for, and why they might be particularly effective. Further, since AI itself has a very distinctive enabling technology, it suggests that this enabling technology, the technology which is used to build AI systems, may be of significant interest in its own right.

1. The AI Toolbox

Contemporary AI is not a hardware technology. It certainly *assumes* a hardware technology. It assumes relatively powerful personal computers with high-quality interactive graphics, "mouse" pointing devices, local networks, color displays, voice output, etc. But that's just the context. Fundamentally, AI is a programming technology. To the extent that AI has anything to contribute to the toolbox of the future, it will be software tools.

AI software technology is characterized by (a few) distinctive programming languages, some quite distinctive programming environments and a variety of very distinctive programming techniques. Although one can argue their merits in the abstract, the case is more compelling in terms of the problem for which these tools were designed. There are some characteristics of the AI programming task which set it apart from conventional programming in rather dramatic ways.

The essence of the AI programming task can be made clear simply by reflecting on the meaning of the phrase "artificial intelligence". "Artificial" is easy enough, that means using machines, which in this context means digital computers. "Intelligence" is somewhat harder. I rather like Minsky's definition of "intelligence" as a property we ascribe to intellectual behavior that we admire but do not understand. Iconoclastic, but effective. For example, a seven year old child thinks long division requires "intelligence" because smart kids can do it and the dumb ones can't. Adults know better, they know what the trick is. But when it comes to natural language understanding, everyone agrees that's "intelligent" because none of us know how that works.

The implication for AI programming of Minsky's unsettling definition is that, if you are building an AI program, you are, *by definition,* building a program that you don't understand. The going in position is that you don't know what you're doing. And there's no reprieve. If you *do* know what you're doing (or if you find out), it isn't AI anymore, it's something else, because it can't possibly involve "intelligence". Some topics have actually been reclassified out of AI, for just this reason. For example, symbolic mathematics was once one of the classic AI problems. However, eventually algorithms were discovered for many of the basic operations which performed much better than the heuristic AI programs could. The field was reclassified. Symbolic mathematics is no longer AI, because we're too good at it.

People who don't know where they're going don't take a lot of baggage. AI programming is invariably done by individuals or very small groups (often of captive graduate students). The reason is clear. If you don't know where you're going, you'll probably start out in the wrong direction. It will therefore be important to be able to change direction quickly. That mitigates against large programming teams, because it just takes too long to turn them around. In any event, the large team would probably be unmanageable, given the lack of a detailed specification, since there would be no effective way to subdivide the problem so they could all work effectively.

None of this would be very important were it not that AI programs tend to be big. Whether the complexity is in the data, as in "knowledge based" systems, or in the program itself, as in "inference-based" systems, the net effect is the same. Given a small implementation team operating under uncertainty about the final design, the result of attacking any large problem is a program that is out of control. Being out of control is standard operating procedure in AI programming. Not because the programmers are bad (in fact, they are usually very good), but because the size and difficulty of the problems that they are dealing with always threaten to overwhelm them. Indeed, in many ways, the research boundary of AI is (perhaps disappointingly) simply a direct reflection of how complicated a program can be sustained by a small group at any given moment in time.

2. *Programming Under Uncertainty*

The effect of this "complexity standoff" is that a tremendous amount of effort has been invested within AI to build programming technologies that would help people cope with the problem of *programming under uncertainty* — programming when you don't know what you're doing. Central to all of these approaches is the observation that, if you don't know what you're doing, you have to find out. The only way to find out is by trying something, by experimenting, by exploring the domain, and seeing what works. Any such *exploratory* orientation places an enormous premium on making it easy to change your mind. Unless you're lucky, you usually don't get the right answer when you try something. Instead, you get the answer, "Not quite right". So you must expect to be in a continual state of tearing your code up in response to the latest failure and rebuilding it until you converge on a solution.

Notice how completely orthogonal this is to conventional programming methodology. There, one begins by writing down the design, checking it until you're sure you got it right, and then locking the coding process into that design so you implement exactly what the design specifies and nothing else. That's absolutely *not* what we want here. We want to decide what we want at the *end* of the implementation process, not at the beginning when we know almost nothing about the problem. That only works when you have perfect foresight, which we most certainly don't have in AI. As a result, most conventional programming methodologies can be (and are, in AI practice) discarded immediately. They don't work. They don't work because their objectives are completely different from the objectives of AI programming. AI programming accepts the need to explore a problem, to find out what works. Conventional programming methodologies try to limit such exploration, by enforcing adherence to a design whose properties are perfectly understood in advance.

The experimentally oriented AI programming technology has three distinct components — languages, environments, and techniques. A detailed technical account of these would be out of place here (it can be found in Sheil, 1983 or Teitelman & Masinter, 1981). However, a brief review serves to illustrate the basic differences in the technology and its sources of leverage.

3. Languages

Computing technologists get very agitated about issues of programming language. From an outsider's point of view, the debates have a fainty theological air, both in the narrowness of the issues debated and the level of emotion which they evoke. Nevertheless, from both points of view, the majestic domination of AI programming by Lisp and Lisp dialects for over 25 years is quite remarkable. Unlike the economic motivation of large amounts of existing code, which kept languages like FORTRAN alive across a similar sweep of computing history, Lisp has survived and prospered in the AI community because it has embedded within it some design choices which are a singularly good match to the AI development style. These issues, technical details aside, are important for what they tell us about the nature of AI programming.

One key principle is to allow the programmer to *defer commitment* as long as possible. A decision that has not been cast into code does not have to be recast when it is changed. Thus, the longer one can carry such a decision implicitly, the better. In some cases, a decision can be deferred until later during program development. In others, it can wait until program compilation (so no code needs to be changed when the decision is changed). And, in some cases, the decision should be deferred all the way until run time (so it can be changed dynamically, while the program is running). For example, it is rarely advantageous for the programmer to predefine the exact usage of storage, and AI languages are as one in providing automatic storage allocation and garbage collection. Unlike most conventional languages, AI languages are untyped (any variable can refer to any value) and generally permit dynamic free variable reference, which makes it easy for the programmer to add new data paths between procedures. One can pass procedures as arguments, construct new ones and return them as results, set a procedure to be the value of a variable, or even redefine a procedure on the fly, while your program is running. These provide ways of deferring commitment about what an operation means. In the limit, using the technique of *object-oriented programming,* one can actually bury the procedures associated with a piece of data inside its value, so that each object carries with it its own ideas about how to respond to certain operations. All of these are techniques for deferring commitment, for making experimentation easier by minimizing how much you have to do when you back out and change your mind.

The other critically important idea in AI languages is the notion of supporting embedded languages and metalanguages. In the case of Lisp, this was enabled by Lisp's simple, easily described within Lisp, representation for Lisp programs. This made it straightforward to write Lisp programs which could manipulate other Lisp programs as data — reading them to determine what they did, making changes to them, creating new code from them. This enabled experimentation on and extension of Lisp itself to take place within Lisp. This was important both technically and sociologically. Sociologically, it bound the language developers and users into a common culture and propagated these languages extension techniques into wider use. Technically, it "closed the loop" and made the progress of language evolution much faster. Although we tend to talk about "Lisp" as if it were one language, it's actually a moving target that has absorbed many ideas from other languages because of this powerful bootstrapping ability.

More important than the effect of embedded language techniques on Lisp itself is the idea of taking an application specific concept and embedding it in Lisp, producing a new language which is crafted directly for the application as it is currently understood. For example, you can take the problem-solving strategy of deterministic backtracking and embed that in Lisp, with its own operations, syntax and semantics, to form a new language (in this case, PLANNER) which is all of Lisp, plus something else which makes it easy to use that particular programming style. This is enormously powerful because it allows you to start *thinking* in those application specific concepts. You have an idea about what's important, you build your idea into the language and now you think in terms of your idea. It's also enormously seductive. You can fall in love with a language; you cannot fall in love with a subroutine, it's just not possible. Language embedding has been used to develop a whole series of Lisp sublanguages (e.g., PLANNER, CONNIVER, QA4, etc.), each of which has been used to explore the AI problem solving paradigms that it encapsulated. Each one demonstrated the enormous power of a language as a carrier of an idea. As a group, they demonstrate the enormous power of Lisp, the meta-language in which successive generations of AI idea languages have been expressed.

4. Environments

Impressive though the exploratory programming languages are, they only speak to half the problem. Their key feature is that they encourage the programmer to defer commitments as long as possible, in the interests of preserving mobility. In some cases, such as automatic storage management, the programming system has to absorb a significant burden from the programmer in order to allow this freedom from commitment. However, more has been allowed than absorbed. Since there is (by assumption) no complete design to work from, the programmer still has the significant problem of keeping track of a large, rapidly evolving system. Further, in the interests of constraint deferral, most of the language features, such as static type checking, that might have supported conventional programming methods have been removed, although we have argued above that these methods would in any event be ineffective. How is the programmer to regain control?

The essential element is an integrated programming environment which uses language-specific programming knowledge to provide exactly those control and bookkeeping functions that are the greatest drain on a programmer during rapid system development. An environment like this can actively assist the programmer

- by deriving and keeping track of information about programs as they change;
- by providing ways to view this information and use it to make systematic transformations to the evolving system;
- by providing active "agents" that can notice events that the programmer should pay attention to, take action on them, propose changes, and automatically take care of routine details (like cleaning up after a session); and
- by supporting ad hoc performance engineering of those parts of the system that have achieved design stability, so that one does not pay a performance penalty for one's initial uncertainty about the design.

As before, we will eschew detailed technical discussion (for which see Teitelman & Masinter, 1981) in favor of some brief illustrations drawn from the Interlisp–D environment (Xerox, 1983). This environment maintains a number of different databases about the programmer's actions and the code that is being developed. One key piece of knowledge is where in the file system to find each program's source code. Thus, if you decide to edit a particular fragment of code, you can ask for it by name (or some other meaningful description like how it is used) and the system will find it, bring it to you (in some appropriate editor) and then put it away when you're done. During a sequence of edits, the system keeps track of what has changed so it can update and recompile the corresponding source files afterwards. The system can read and analyze user programs so that it can respond to commands like, "Find me all the places where I've done such and such." Note that all of these are critically dependent on the ability of the programming support tools to look at the user's code as it is written or changed to derive the information on which all this depends. So it is not surprising that these tools have gained their greatest sophistication in Lisp based environments which support this notion fully.

In a modern AI programming system, these knowledge based facilities all have elaborate interactive graphic interfaces which provide specialized viewers and information management systems to support programming in their particular language. What's important though, is not the viewers, but the information that's behind them. The key enabling feature is that the system understands what's happening, so it can act for the programmer, and that the programming tools are integrated so that they share information. It is this understanding and cooperation that allows the environment to help manage the program complexity that is inevitable during fast track development.

5. Programming Techniques

The third component of AI software technology is a wide range of programming techniques that have been developed to facilitate AI programming. Some of these have been, for one reason or another, intimately connected to particular languages or environments, so we have encountered them already, e.g., language embedding and automatic backtracking. Others of note would include data-directed programming, rule-based systems, constraint propagation, and various systems for knowledge representation (for a complete treatment, see Charniak et al., 1980).

Oddly enough, despite the fact that they were developed in AI contexts, most of these techniques have considerable *general* programming applicability. Although the problem to which they are directed may have arisen first, or more acutely, in an AI context, there is little about any of them which is particularly idiosyncratic to AI applications. For example, rule

based systems, although central to the "expert system" style of AI application, are a general programming technique for structuring large systems
so that small additions can be made incrementally without having to understand the entire control structure. This is a very powerful programming
idea. It has nothing to do with intelligence, per se. It's got to do with managing large programs that need frequent, local changes.

6. Applying the AI Toolbox

Given this brief review of AI software technology, let's return to the question
of why this might be of interest to anyone outside the AI development
community. We have seen that the AI programming tools were designed
for a particular style of programming, which I referred to as *programming
under uncertainty*. In this situation, where there is significant design insta-
bility and limited ability to predict how effective any given solution will
be, the AI programming tools have shown themselves to be remarkably
effective in helping the programmer to explore the design space rapidly
and converge on an appropriate design.

The relevance of this technology outside of AI should be clear — coping
with uncertainty is a completely general and increasingly common programming problem. Looking at the data processing applications which remain for the 1980s, one finds case after case which, although not AI, comprise classic instances of programming under uncertainty. The easy
applications, the places where there is a well-defined, easily separated,
procedure that can be automated routinely, have all been done. Increasingly,
we are confronted with situations where there isn't a well-defined procedure,
where no one knows what's really needed, where the existing data processing staffs have been turned away, usually with a failure in hand. We
have situations where nobody knows how to start. These are ideal matches
to the strengths of AI software technology. That technology won't make
the exogenous factors that contribute to the uncertainty disappear, but it
at least holds out the promise of being able to adapt to them fast enough
so that the uncertainty can be mapped and the range of system design
possibilities can be understood.

The effectiveness of AI software technology as an application development vehicle is currently being discovered in the support environments
being produced to package the first AI applications. An AI application (e.g.,
an expert advisor on subject X) is of little use in isolation. The AI functionality must be embedded within a system that provides or interfaces to
a significant information management function (e.g., a database, analysis
or filing system) for the application area. Quite often, however, interactive
information utilities have never been constructed for the intended users

of these AI systems. Equally frequently, the information utilities currently in use are ponderous compromises limited by the inability of a development team using conventional technology to track the needs of their users. As a result, the supporting information environments for many of the first AI applications are likely to have a revolutionary impact on their users. And since the ability to build highly customized information management systems is itself an application opportunity of enormous leverage, it will not take long for the application development power of AI's software technology will be recognized and separately deployed.

In the longer term, the AI development technology should spread even wider. Applications clearly lie along a *spectrum* of uncertainty, between the very well understood ones (the ones that are all done now) and the extremely ill-understood ones (which are not done because no one knows what to do about them). AI occupies the latter extreme of that spectrum. Conventional data processing occupies the other end. Until recently, there has only been one cost-effective application development technology — one that assumes that everything is fixed and decided and certain. As a result, the application spectrum was dichotomized and everything that was considered doable (which did *not* include AI) was deemed to be completely understood because the application methodology demanded it. The availability of a cost effective technology for dealing with design uncertainty removes the need for any such artificial barrier and suggests that we should revisit many applications with which we have had little or no success using conventional development methods. For example, an enormous variety of specialized, professional information support systems now seem to be realizable given AI's software development tools.

As long as "intelligent" continues to imply "undefined", artificial intelligence will continue to push the art of programming as exploration. Thus, however much specific AI techniques become incorporated into standard software practice, AI will continue as a key source of software technologies for our most "way out" projects. Thus, not only does it seem reasonable to embrace the description of AI as the "tools of the future", but it appears that it may hold true for quite some time.

7. References

Charniak, E. *et al. Artificial intelligence programming.* Hillsdale, NJ: Lawrence Erlbaum, 1980.

Sheil, B. Power tools for programmers. *Datamation,* February 1983.

Teitelman, W., & Masinter, L. The Interlisp programming environment. *Computer,* April 1981, *14*(4) 25–34.

Xerox. *The Interlisp reference manual.* Pasadena, CA: Xerox Special Information Systems, 1983.

Integrating Vision and Touch for Robotics Applications

Ruzena Bajcsy
COMPUTER AND INFORMATION SCIENCE
DEPARTMENT
UNIVERSITY OF PENNSYLVANIA
PHILADELPHIA, PA 19104

What are the prospects for integrated vision-touch systems over, say, the next five years?

We begin by reviewing current multifingered hands, equipped with currently available tactile sensors, as data acquisition devices. What can they do? What is the range of possible applications? Next we look at currently available visual data acquisition systems and their uses. The discussion considers such questions as how to tell shape from touch, and the kinds of algorithms that can be defined given currently available tactile and visual capabilities.

Next, we provide a critique of these currently available capabilities, focussing in particular upon their lack of flexibility, and on the need for visual-tactile integration. We describe our own work toward such integrated systems.

Finally, we examine human visual-tactile processing to see what it suggests about how machine processing might be improved. In particular, we try to distinguish those human capabilities that more sophisticated robotics applications would appear to require from those that do not seem essential, and suggest how the corresponding machine capabilities might be utilized.

1. Introduction and Motivation

During the last decade advances have been made towards a realization of Robots. The news media are full of reports on the expansion of use of Robots in industry, and the justification of this use in terms of the increase of productivity, quality control, etc.

The Robots currently available are by and large unsophisticated and simple-minded. They work in a very constrained and well-controlled environment with tasks that are restricted and deterministic. Examples of this approach are paint spraying, welding, pick and place operations and their like. By this we mean control of illumination, and using only a small number

297

of different objects and/or parts well specified in their shapes, their materials and other physical properties. In addition, the exact sequencing of operations on the assembly line and the procedures for manipulation of these objects is usually known. Under these restrictions it is relatively easy to design an automaton (deterministic by definition) which will perform satisfactorily, and one does not need a Robot. Indeed frequently this is the case in today's factories. The difference between the automated machines and today's Robots is the computer, which allows a robot to be reprogrammable and hence more flexible than a fixed automated machine. This capability will enable even small companies with a small volume of different products to use the reprogrammable robot to perform various tasks. It is the **flexibilibity** which is so impressive about humans, as opposed to machines, which we are striving towards in our quest for **future Robots.** This flexibility comes in part from the fact that people use more than *one sensory input* during their recognition procedure; hence our interest in the integration of *multisensors.*

In this chapter we shall concentrate on a subset of capabilities of a flexible future robot, those required to recognize three-dimensional shapes using visual and tactile sensors. We shall assume that each sensor is processed by an individual processing module; then the results of the two processes are integrated during the recognition process. What follows is a discussion of existing vision and tactile modules, what each can do independently, and then their integration in a recognition task.

2. Vision Module

The vision module is composed of three basic parts:

1. The data acquisition part
2. The feature extraction part
3. The recognition part

Before we launch into the details of each of the parts, perhaps it will be useful to formulate the overall task of the vision module in a more informal way. First, one has to recognize that the peculiarity about visual data acquisition is that it takes three-dimensional information and projects it into two-dimensional information. During this projection there is a loss of information. Hence the most difficult task of the vision module is how to reconstruct the three-dimensional information from these two dimensional projections. In addition to the loss of information due to projection there is a loss of information due to illumination, reflectance properties of the object, and the inadequacy of the imaging device.

The feature extraction process is determined by the different representations and primitives that are being considered. Finally, the recognition process will deal with problems of how much a priori knowledge will be needed, given the data, in order to recognize different objects.

2.1. Three-Dimensional Data Acquisition

The problem addressed in this section can be stated as follows: the two-dimensional digitized pictures are some representations of the illuminated 3–D real world. In order to be able to interpret these pictures in terms of the 3–D real world, we need to know what are the transformations that took place between the real world and its picture or pictures. In order to identify these transformations we have to state a few restrictive assumptions. First, let us restrict ourselves to a real world where each object has some geometric form, e.g., the Euclidean geometry holds. Secondly, we assume that illumination of the world is normal; that is, the laws of optics and radiometry hold. Thirdly, the interpretations of the pictures, the labels that we put on the objects, are defined purely from the visual properties that we can depict. Finally, we have to develop various procedures for transforming the 2–D pictorial data into 3–D surface and volume data necessary for the final scene analysis.

The psychologists tell us that the human visual system has various cues for perceiving the 3–D world, though the images on the retina are 2–D projections of this 3–D world (Rock, 1975). They divide these cues into two kinds: monocular and binocular, depending on whether one uses one image or two images simultaneously.
The monocular cues are:

- linear and size perspective
- interposition
- shading and shadow
- texture gradient
- aerial gradient
- gradient of details
- accommodation of lens-focusing for objects at various distances

The binocular depth cues are:

- stereo
- motion parallax
- converging of the eyes—depending on the distance of the object

Interestingly enough, only the accommodation of lens and the converging of the eyes give us a true measure of distance. All the other cues offer only a relative measure of the distance.

2.1.1. Monocular Depth Cues. The most important feature in monocular depth cues is a continuous gradient (Gibson, 1950) of various features in the image which is interpreted as a change of the reflected light; the aerial gradient is a continuous shift in color, green to blue. The linear and size perspective, the texture gradient, and gradient of detail can be explained by laws of perspective geometry. Interposition as a cue is based on the assumption of continuity of the 3–D world. Horn (1970, 1979) has thoroughly analyzed the effect of illumination, reflectance, and shading with respect to the geometry of surfaces. The result of Horn's computation is a representation of surface normals, as shown in Figure 1b. The original is in Figure 1a.

While this method is able to compute relatively detailed surface information, the aerial gradient implemented by Sloan (1977) can only make global statements about the object being closer or farther.

Historically, linear perspective was used first in computer vision by Roberts (1963). Roberts has used a polyhedral world as the testbed for writing recognition programs. Given a model in 3–D space and a picture in 2–D space, we can find the best match for 2–D into 3–D by the following matrix transformation:

$$H = R * P$$

where P is a perspective transformation, R is the real space transformation of the model, and H is the transformation which takes the model points (in 3–D) into picture points (in 2–D). The underlying assumptions here are

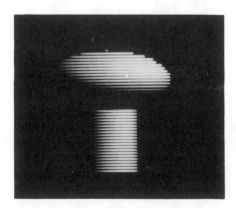

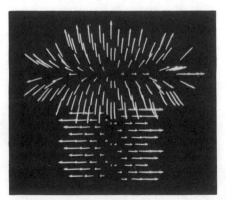

Figure 1a. Original.

Figure 1b. Representation of Surface Normals (from Horn, 1970, 1979).

that the picture is in fact a perspective view of the real world, that the objects shown in the picture could be described by means of one or more transformations of known models, and that all objects were supported by others or by a ground plane. The transformations were restricted to rotation, translation and size change. Then from a single picture each object which has four or more points showing can be described in terms of the models and positioned in a three-dimensional space. The scale of this space in feet can be determined if the distance of the camera can be supplied.

Texture gradient has been recognized as one of the most important monocular depth cues by many psychologists (Gibson, 1950). The gradient of texture will be due to the perceived systematic variation of size of texture elements, and the density and other measures of projected surface texture. The perspective projection has two effects in a textured surface: scaling and foreshortening. Scaling is due to distance; foreshortening is due to surface orientation.

In computer vision, texture gradient was used for the first time by the author and her students (Bajcsy & Lieberman, 1976; Lieberman, 1974). The assumption was made that the analyzed surface is planar and not tilted. We have measured the smallest gradient and the largest gradient. This allowed us to detect the relative distance. From the assumption about planarity of the surface we were able to detect the surface orientation. Stevens (1979) has re-examined the question of texture gradient and devised procedures to compute surface orientation for nonplanar surfaces, as well. Witkin (1980) has continued in the line of research of Stevens and has shown how to estimate surface orientation from contours of projected shadows.

2.1.2. Binocular Depth Cues. So far we have discussed the monocular depth cues and their computer implementation. Now we turn to the problem of binocular depth cues. To recapitulate, they are: stereo, motion parallax, and the convergence of the two eyes. In this section we shall deal with only the first two, since the convergence of the two eyes is a very different process. Both the stereo and motion parallax share very similar computational processes which can be divided into three subprocesses:

1. Production of a camera model; the position and orientation of the camera in 3–D space.
2. Generation of matching point-pairs; loci of corresponding features in the two pictures.
3. Computation of the point in 3–D space for each point pair.

One of the first computer stereo programs was written by Hannah (1974) and Levine, O'Handley, and Yagi (1973). Julesz (1962) studied stereo perception by generating random stereo dot patterns. From this beginning the hardest problem in stereo computation and motion parallax is still the generation of matching point pairs. The workers in this area have used different features for matching, such as edges and intensities (Baker, 1981), or some similar areas (Gennery, 1979), or zero crossings (which are special kinds of edges) (Grimson, 1980; Marr & Poggio, 1979). Barnard and Fischler (1982) and Thompson (1980) have applied relaxation techniques for the matching process, based on continuity of real world surfaces. A very good survey of recent work on stereo computation is Barnard and Fischler (1982). The above cited work shows that stereo and/or recovery of 3-D information from motion parallax is feasible under certain constraints; however, the computation is quite expensive.

It is a well accepted fact, among the researchers dealing with the computation of binocular depth cues, that once the appropriate choice of feature points is made and the matching process is handled, computation of the disparity map and the points in 3–D space is trivial.

2.1.3. Three-Dimensional Object Representation and Feature Extraction. Given that we have scanned the 3–D object completely (from all angles), what are the suitable representations, or parametrizations, or data reduction mechanisms for purposes such as: storing, recognition and description. Geometrically, a three-dimensional object can be described by its surface characteristics, including edges and corners, as well as by its volume characteristics, especially its symmetry surfaces and symmetry axis. From the point of view of topology, additional properties come into focus, such as: connectedness, number of holes, handles, etc.

Psychologists, as well as computer vision researchers, must also be concerned with the problem of two coordinate systems: one in which the data is acquired, called *observer-centered,* and one in which it is finally represented, which is more associated with the object, and hence called the *object-centered.* Notice that the observer centered representation contains only a partial view (at best half of the hemisphere) while the object centered representation will encompass the complete information (composed from many partial views) about the object. This is the crucial difference between these two representations, rather than the actual coordinate system.

2.1.4. Object-Centered Representation. By object-centered representation we mean a representation which describes the object independent of the observer viewing point; hence, it has to be complete and rotationally invariant. The corresponding coordinate system could be either centered inside the object, e.g., in the center of the mass of the object, or at any reference point in the 3–D world, independent of the position of the observer. Similarly the coordinate system could be, but need not necessarily be, aligned with some major axis of the object. Various researchers have proposed different primitives for object-centered representations, such as symmetric axis (Blum, 1967; O'Rourke & Badler, 1979), generalized cylinders (Binford, 1971; Nevatia, 1974; Soroka, 1979) and touching spheres (Mohr & Bajcsy, 1983).

2.1.5. Observer-Centered Representation. In this section we wish to consider the problems not only of partial view, but also of the surface representation of a 3–D object. Again the main problem is what primitives one should choose. Assuming that we have available the 3–D coordinates for each surface point, we can compute the surface normals for all the points. From this information one can consider: an algebraic method, where all points satisfy some equation $F(x,y,z) = 0$; and parametric representation, where a surface is traced out by three bivariate functions $X(u,v)$, $Y(u,v)$ and $Z(u,v)$. The surfaces represented by the algebraic method as well as by parametric representation could be planar or polynomial or in general of any form. In practice, however, one is restricted to planes and simple polynomials. More complex surfaces are then approximated by either planar or quadratic patches (Dane, 1982) or by spline techniques. These techniques are well known from computer graphics, and computer aided design (Foley & Van Dam, 1982). The hardest task in surface analysis and its recognition is to recognize the junctions. At this point it is not clear in how much detail the junctions should be represented and described. Clearly for computer-aided design this question is of more importance than for a general purpose vision system.

2.3. Recognition Strategies

The problem we wish to investigate in this section can be formulated in most general terms as follows: What is the tradeoff between the amount of measured data and the a priori knowledge in order to recognize the object? If we have complete knowledge about the object, and the number of possible objects occurring in the world is small, then clearly with very few measurements we can recognize the object. On the other hand, if we have almost infinite time to measure the object then one does not need to know very much about it and one can come up with rather detailed description. The only a priori knowledge that will be used will be the vocabulary of the chosen primitives. We feel that it is of great importance to understand this tradeoff because it determines the representation as well as the sequencing of actions—including data acquisition and feature extraction—that are necessary for recognition. The strategy generated for this purpose naturally will be different from object to object and from situation to situation. This complexity can be characterized in terms of geometry (Tzikos, 1983).

Figure 2a.

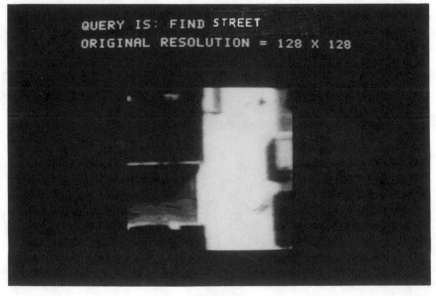

Figure 2b.

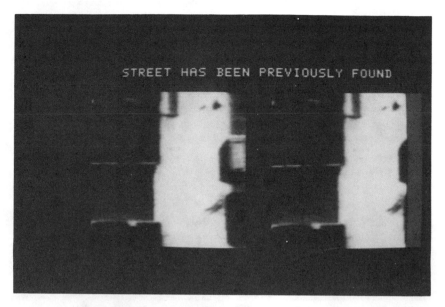

Figure 2c.

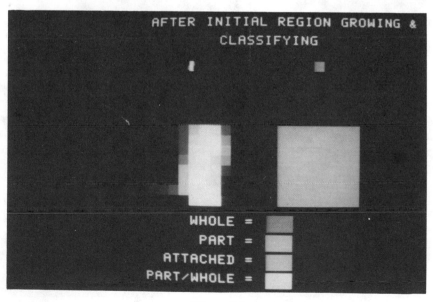

Figure 2d.

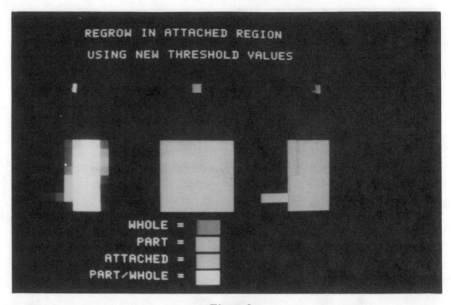

Figure 2e.

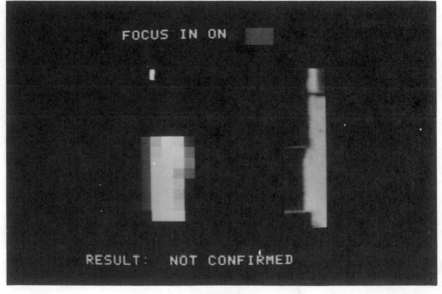

Figure 2f.

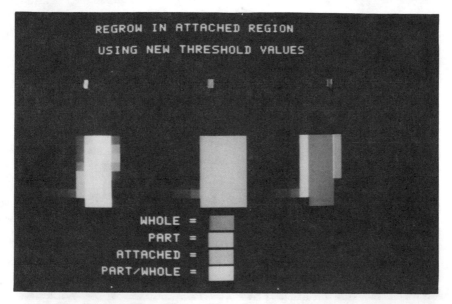

Figure 2g.

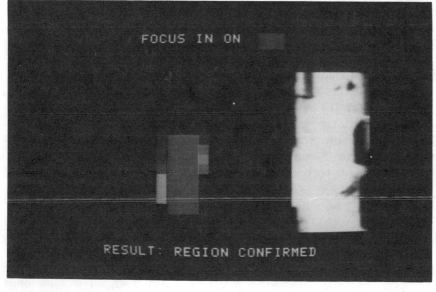

Figure 2h.

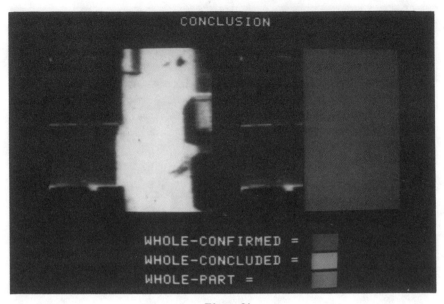

Figure 2i.

Another aspect of recognition strategy is the necessary or desired detail of descriptions. This point calls for a hierarchical approach, where the hierarchy corresponds to the details of the shape of the object. In the past, several authors have proposed hierarchical structures (Hanson & Riseman, 1975; Mishihara, 1978; Tanimoto, 1975; Uhr, 1972). We have applied this strategy to aerial scenes (Bajcsy & Rosenthal, 1980). Some examples of this work are shown in Figures 2a–i. This organization helped us save not only search time in matching, but also in visual processing, since we attended to details only when it was necessary for clarification in decision making.

3. The Tactile Module

Just as the two eyes equipped with all the processing power described above deliver visual information about the world, the hand, with its joints and skin, can be viewed as a data acquisition system for tactile and kinematic information. Under the tactile module we include tactile sensation from the skin, and kinematic information coming from position and force sensations based on the joints of the fingers and the wrist. As in the previous section, there are three parts to this module:

1. data acquisition
2. representation of the tactile and kinematic information
3. recognition

Before going into details, we must point out that research in tactile sensation is far less developed both in psychology and in the Robotics literature. The psychological literature maintains that vision is the superior modality, although this is now beginning to be questioned. In any case, for a flexible robot there is need to have more than just one sensory input, especially if one considers grasping situations, and situations where the illumination may be inadequate.

3.1.1. The Data Acquisition System. The data acquisition system is composed of two different systems. One is the hand, which is composed of multijointed fingers with position and force sensors in the joints. The other is the skin, which is an array of pressure-sensitive sensors distributed over pads which are placed between the joints.

3.1.2. Current Mechanical Hands. There are several aspects of the design of a mechanical hand that we shall review here. The global task is to produce a highly versatile hand with a minimum of moving parts, a dependable drive system, and an optimum number of degrees of freedom (D.O.F). The hand will be an assembly of motors, sensors and mechanisms (fingers).

We start with the fingers: Loosely speaking fingers should have one or more bending sections (links). Each finger link should be a component of a closed linkage connected with other links via joints which can "drive" or rotate the link. Fingers should be mechanically identical and substantially attached to the hand's base. The fingers should be able to approach, contact, or pass one another during prehensile operation. There are two basic joints and two common combination joints (Roth, 1975): revolute (R) and prismatic (P). A cylindrical joint (C) is a combination of a coaxial R and P joint. A spherical joint (S) is formed by a ball and a socket and is denoted as RRR joint or three noncoplanar independently powered R joints. The linkages are mechanisms that have been studied in the past, for example (Paul, 1979). The mechanisms studied to bend the fingers are cross four-bar chains, miniature compound pulleys, four-bar chains with an expanding link (for example, alternate screw threads), and tension cables (without pulleys). Cross-bar chains are dependable, easily built, and can transmit an angular displacement, "w", from one chain to the next, producing a rotation "w + dw" in a continuously compounding manner.

Miniature compound pulleys develop a high mechanical advantage and allow the finger links to bend through large angles. However, straightening a bent finger is difficult.

Expanding link four-bar chains can be driven open and closed easily with a high mechanical advantage and reasonably low losses. This is a complex drive mechanism which requires a translating torque transmission or an expanding piston-cylinder motor in each link.

Tension cables flex a finger or its joints by simple direct contact, and very little space is required. To achieve high prehensive forces, high cable tension must be maintained.

The number of optimum degrees of freedom (D.O.F) is defined by the capability of a manipulator to grasp all the basic geometrical shapes from any aspect with a minimum of external control inputs. The basic shapes under consideration are: rectangular and triangular prisms; spheres, and cylinders; screws, and rods and their like.

In addition to the requirement of approachability of the mechanical hand from any arbitrary position during grasping, it is necessary that the hand constrain the grasped object. Hence, the optimum number of links per finger would be the minimum number necessary to achieve this constraint.

There have been three different applications that motivated hand development: prosthetics, remote manipulation and robotics.

Historically, the first mechanical hand seems to have evolved as a prosthetic device replacing a lost hand. Indirectly we learned that in 1509 a gripping device was designed for a knight who lost his hand in battle (Bajcsy & Rosenthal, 1980). In 1962 we heard for the first time about the Belgrade hand (Salisbury, 1982; Tomovic, 1960). Later consequent reports came from the same laboratory (Jaksic, 1969; Tomovic, 1969) and the French laboratory (Clot, Rabishong, Peruchon, & Falipou, 1975; Rabishong, Stojiljkovic, & Peruchon, 1969) and the USA (Crossley, et al., 1977; Skinner, 1975; Sword & Hill, 1973). Recently at the University of Utah an elaborated prosthetic hand controlled by myoelectrical impulses has been reported (Jacobson, 1982). Almost without exception prosthetic hands have been designed to simply grasp objects, and so the dexterity of the artificial hand in itself is not that important. The control by the user-amputee comes from his/her visual source, which guides the grasping process.

The second related field of applications of artificial hands has been in remote manipulations. These are the situations where one needs to work with hazardous materials such as occur in a nuclear environment, or at a distance as undersea or in outer space. Here the control is usually by a person in a master-slave or teleoperator mode. Examples of these efforts are: The JPL and others effort for NASA (Bejczy, 1977; Skinner, 1975) and for robotics (Hill, McGovern, & Sword, 1973).

In 1977, and again in 1979 and 1982 a versatile finger system was reported on (Okada, 1979, 1982; Okada & Tsuchiya, 1977). Okada's system is the most complex amongst that generation of hands. It has three fingers with 11 D.O.F. Two fingers have four revolute joints and the third finger has three revolute joints. The joints are driven with wire which runs inside the finger in coil-like hoses. Thus it is unnecessary to set relaying points for guiding the wire at the joints of the hand, and the path of power transmission could be selected freely. It was found that fingers with circular cross section were best for 3–D motion. A 17 mm in diameter free-cutting brass rod was cylindrically bored. The tip of each finger was truncated at a slope of 30 degrees. The bending and expanding of each joint was accommodated for a maximum angle of 45 degrees bending in and out. Thus finger work space is more extensive than that of human beings. The hand is capable of rotating spheres and turning and twisting rods in the air. The control of the fingers is in a combination of position and torque control. Torque and position are both controlled by the hardware servo.

A multipurpose mechanical hand for industrial robots has been developed by Rovetta (1977; Rovetta & Casarico, 1978). Since every finger is independent of the others, n of the fingers may be used by means of a selector. By successive sequences of the various fingers, different grasping and assembling operations may be carried out.

A curious design has emerged from another Italian laboratory led by Nerozzi & Vassura (1980), where the primary concern is how to grasp and handle delicate agricultural products like pears, peaches, etc.

Finally, the most advanced hand published so far is that of Salisbury, shown in Figure 3. This hand has three fingers and each finger has three joints.

3.1.3. Sensors. Manipulator sensors are of three types, according to Kinoshita, Aida, and Mori (1971): force, inertial and tactile sensors. The primary interest of this paper is in tactile sensation. Force sensing is as it may serve as a detector of contact between the hand and examined object, or as a detector of torque force. Inertial sensing will be ignored here completely.

Existing force-sensing devices can be classified into one of three categories (Shimano, 1978) according to their placement relative to the manipulator with which they are used. Force sensors are mounted on the joints or at the end of the manipulators (at the hand and/or fingers) or on the support platform of the manipulator. For the best accuracy in sensing the load, the most desirable place for the sensor is the very end of the manipulator. However, that requires a miniature, lightweight implementation of the sensor and the associated electronics. A similar rule applies for placing the servomotors. If motors are closer to the sensors which provide the feedback, then control is simpler than otherwise. At the same time, the motors add weight to each link, which complicates the control issue. As usual, the actual implementation is a compromise and typically the sensory devices are distributed throughout the manipulator. Sensing torque is the oldest mode of force sensing at each manipulator joint. It detects

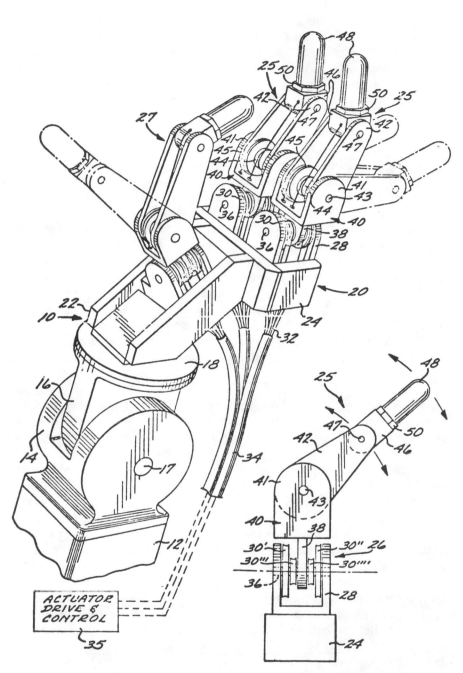

Figure 3. Stanford/JPL Hand. *The hand is shown on a Unimate 600 manipulator.*

not only forces and moment at the joint, but also at other points on the manipulator. The drawback is the uncertainty in predicting the friction, and joint damping is sufficient to preclude the possibility of accurately detecting small hand forces. Over the years the wrist has evolved as the most desirable place to put force sensing (Flatau, 1973; Scheinman, 1969; Whitney, 1977). As an example we shall describe the "Scheinman wrist". Fashioned like a Maltese cross, the wrist motions are measured by a series of semiconductor strain gauges mounted on the cross webbings. There is one gauge on each side of four webbings, hence 16 gauges. A similar system at MIT, with different geometry, was used for some delicate peg-in-hole assembly tasks (Inoue, 1974, Shah, 1972; Silver, 1973).

Interest in touch sensing for flexible Robots dates back to the mid-1960s (Sutro & Kilmer, 1969). Already at that time researchers recognized the need for contact feedback, especially in remote manipulation. The problem was and still is the lack of adequate tactile sensors. Binford (1972) presented a coherent review of the state of the art of sensors as it was in 1972.

To our knowledge, Kinoshita, Aida and Mori in 1973 were the first to report on pattern classification of a grasped object by an artificial hand. The artificial hand had an array of 20 pressure sensitive elements (5 rows by 4 columns) and the shape of the surface was identified by one single grasp. Similar tactile array arrangements placed on the hand were reported by Hill, McGovern, and Sword in 1973. Bejczy (1978) describes a tactile array of 32 elements, and Clot and Falipou (1978) and Briot (1977) an artificial skin of variable resolution (10 by 10 and 24 by 24 elements). The difference among these reports is in what kind of physical principle the sensors are based. The most commonly used sensors are based on:

1. strain gauges
2. microswitches
3. piezoelectric changes
4. magnetostriction
5. change of resistance with compression (conductive rubber)
6. piezoresistance

Harmon very recently has published several reports on studies of various industrial tasks and where they would need a tactile device (1980, 1982a, 1982b). Also from these studies a specification sheet follows for the most desirable properties of such a sensor. These are summarized below:

1. The touch sensing device is assumed to be an array of 10 by 10 force sensing elements on a one square inch flexible surface, much like a human finger tip. Finer resolution may be desirable but not essential for many tasks.

2. Each element should have a response time of 1–10 ms, preferably 1ms.
3. Threshold sensitivity for the element ought to be 1g, the upper limit of the force range being 1000g.
4. The elements need not be linear, but they must have low hysteresis.
5. This skin-like sensing material has to be robust, standing up well to harsh industrial environments.

These are of course stiff requirements, and the prediction is that perhaps 5 to 10 years of research may lead to such a sensor. Briot (1977) has developed a touch sensor based on changing electrical resistance of contact between an electrode and a conductive silicone rubber. This sensor is very similar to the one developed by Purbrick (1980) and later advanced by Hillis (1982), both from MIT. The advantages of the MIT sensor are finer spatial resolution (256 sensors fit on the tip of a finger) and fewer wires. For a matrix of NxM, the MIT sensor requires only $N + M$ wires as opposed to the French artificial skin, which requires as many wires as there are sensors.

A true solution to the wire problem has been put forward by Raibert and Tanner (1982). His approach is based on VLSI technology. This technology allows coupling the sensor with the computer, which is necessary for processing the signal, and thus reducing the information to fewer bits than the raw signal, hence reducing the number of wires.

Overall Evaluation of the Hand and the Tactile Sensors. From the studies of Harmon and others, it is clear that touch is a useful and desirable modality for robots, in particular for grasping, fitting parts, gathering spatial information in a poorly illuminated environment, and discerning the hardness, liquidity and elasticity of materials. In research laboratories there are available multifingered and multijointed hands, though these are not yet commercially available. The hands are usually equipped with force and position sensors. The tactile arrays are put on tongs only, rather than on multifingered hands. The biggest drawbacks of the current tactile devices are coarse spatial resolution, lack of robustness, and at the same time lack of sensitivity. However, even if better devices were available, we still would need work in data processing and development of algorithms for interpretation of the tactile data. An example of a finger in my laboratory equipped with 123 tactile sensors is shown in Figure 4.

4. Integration of the Visual and Tactile Information

Before one can think of integration of two or more modalities one needs first to understand the capabilities and the weaknesses of each independently. That was the reason we covered vision and touch separately. There

Figure 4.

are, however, additional needs that follow from the integration process. Those are: the form of representation of the information that is collected from vision and touch, and the control structure that will be employed using the two modalities for recognition purposes. Hence we view the integration process as a specially designed representation and control structure that uses two modalities: vision and touch.

Comparison between vision and touch suggests that while the visual and the tactile sensors both have a 2–D receptive field, they do measure different physical properties of the world, i.e., reflectance and pressure. The eye and hand movement both provide 3–D information, but the eye has a higher spatial resolution than the tactile sensor. These similarities and differences of course imply different requirements for the knowledge representation and the control structure needed for recognition if the two modalities are available.

4.1. *Representation needed for integration*

What is different about the representation of the visual and tactile information if treated independently versus if they are going to be used in an
integrated system? As contrasted to the situation when one has only one
kind of a sensor, all the knowledge needed to supplement the inadequacy
of one sensor only can be now assumed to be provided by the other sensor.
Hence the fixed a priori knowledge necessary for successful recognition
will be much less since more and greater variety of measurements can be
made. What we will have to provide is a mechanism (slots) for gathering
this information and extracting the appropriate features. What kind of
knowledge do we need?

1. Knowledge about the individual sensor
2. Signal processing procedures which will extract low level properties:
   edges, regions, texture descriptions, feature descriptions like corners, points of extrema, surface normals, elementary surface descriptions, hardness of material, etc.
3. Geometric models of three-dimensional objects, i.e., the classes of
   objects to be recognized.

4.2. *Control structure needed for integration*

There are three different modes, each of which will imply a different control
strategy. One is initiated by a general request: "Describe every object that
is there". The second is when the request is object specific: "Find the chair
whose seat is made of green leather, has four legs and whose support material is iron painted gray." Finally, the third mode is somewhere in between
these two extremes, and can be characterized by the following request: "Is
there a chair which is free to sit down in?"

Each mode has three stages: the initialization, when from the request
either the sensors become activated to produce an initial scan of the scene
and then make hypotheses on what could be there; or, the knowledge base
generates a list of attributes that the sensors and the feature extraction
processes should generate. The second stage is sensory processing and
interaction with the knowledge base, making hypotheses and verifying the
intermediate results. Finally, the last stage is producing the results.

In general, the vision sensor is used with higher priority than touch.
It provides a more global view of the scene and it is used in the beginning
to make hypotheses. Touch is then used to verify an hypothesis. In the
case when visual recognition leads to multiple and/or ambiguous interpretations, touch will be used for resolving this ambiguity. In the case of
poor illumination, the hand will be used first for making hypotheses or just

simply taking actions. For example, if one walks into a dark room, the first thing one does is search by touch for the switch to turn on the light. Another example is if one looks for an object in a bag, one first feels the object, then pulls it out for visual examination.

Very little concrete work has been done in this area. Recently we have started such work by implementing a simple scenario where we used the visual information predicting the planar surfaces of a prism and then used the finger (as opposed to a hand) for verifying the visible surfaces as well as detecting the nonvisible surfaces. Clearly this is just a mere beginning.

4.3. Human visual-tactile processing

As in studies of integrated systems of vision and touch for robots, the integration principles and/or requirements will follow from consideration of the two modalities and then how they are used in parallel and/or complementary fashion for recognition purposes. There are two aspects of both modalities that need to be discussed separately: the quality of the sensors and the application strategy.

There is great similarity between the visual and tactile sensors. They both have a two-dimensional receptive field, though with different spatial resolution. Temporal resolution, similarly, is used to compensate for image fading.

Perhaps the greatest difference between the two sensors is in the development of the EGOCENTER. The contrast is displayed below:

VISUAL EGOCENTER is the center of phenomenal space. It is a particular location within the physical space depending on the head's position. If the head is stationary and the eye is free to move, then the monocular egocenter coincides with the eye's center of rotation. If the head moves then the center will be where the head moves.

HAPTIC EGOCENTER is not fixed in position. It varies with the part of the body that is used in making the measurements.
The important result here is the frontal plane hypothesis: the orientation relative to the torso is what determines at least one axis of the haptic center.

The next comparison will be based on strategies employed by vision and touch in gathering and interpreting the data. Here we view the sensors as ACTIVE SENSORS which investigate and measure the objects and eventually reconstruct them.

The Investigation Process

Visual tracking

The field of "instantaneous" vision is wider than that of touch. The amount of information immediately available is quite large. As the field of instantaneous vision is restricted, the pattern of eye movement approximates the outline of the perceived object.

Tactile feeling

In the feeling process the contour is, as it were, "unwound" into an ordered sequence of tactile signals. Each finger has a different part in the process of "unwinding" the contour and it traces out only a particular section.

Measurement

Both modalities measure the SIZE of the objects and their parts. There is a direct relationship between the distance and the time the eye or hand must traverse to gather the information. There is a discontinuity in saccadic eye movements that is similar to the discontinuity in hand movements during the information acquisition process.

The Reconstruction Process

Studies of tactile feeling of planar figures show that the dynamics of the interaction of fingers are determined by the shape of the outline of the figure. For example, two fingers are used for tracing a straight line, three or four for broken lines or curves. If only one finger is used then the outline is followed, with occasional visits crossing over from one side to another.

In Vision the reconstruction function does not stand out so clearly. Experiments show that the eyes do not follow the contour of the object, especially under short exposure. With slow exposure, they, like the hand, find reconstructive points of the object.

The reckoning points are the construction points. The feeling hand does not simply reproduce the contour of the object in a continuous fashion, but rather it creates a true representation of it.

In conclusion, it appears that eye movements are far less dependent on the shape of the object than hand movement. They both find the reckoning off points. The eye, however, has more freedom in choosing the path of visual examination and hence it gives the impression of disorganization.

5. Conclusion

The above review represents the state of the art in computer vision and tactile information processing as it is in the **research** and **academic** communities. Commercially available vision modules are by and large more primitive than those described in this paper. The vision modules usually use one camera (no binocular vision) with orthogonal projection (no perspective distortion) and only black and white colors as opposed to gray-valued pictures. The software available can locate a small variety of parts if they are isolated, and most recently even if they are touching or partially occluded.

The tactile sensors are even less developed than the vision modules. The only one currently available is at the Lord Corp., which has an array of 8 by 8 pressure sensitive sensors on a pad. There is no software available to go with the sensor.

The most important difference in the flexibility of people versus robots is in the transformation from sensory signal onto some chunks or segments of information that are labeled. This transformation process is commonly referred to in the literature as a segmentation process and it is currently one of the most difficult and unresolved problems.

How to choose the proper segmentation criterion depends on many variables. Some of those we know, but some have not yet been identified.

One of the most important a priori variables that determines the segmentation process is the choice of primitives, i.e., the elementary discrete units into which we assume the whole scene can be decomposed. What the form of this primitive or primitives is is commonly referred to as the **Representation problem.** While researchers have investigated many different representational schema for visual objects, very little has been done for tactile information. One does not need to be an expert to see that different objects and scenes require different representations.

The next important question follows, and that is: How to use multiple representations; again we are faced with the **question** of **flexibility,** in particular with flexibility in recognition strategies.

The problem of recognition strategies can be formulated as follows: Given a set of different representations, how to decompose the data into those representations which will be "best" suited for the desired interpretation and/or output.

Here we encounter the next variable influencing the segmentation and that is **context.** Questions like how do we recognize context, and how many multiple representations do we generate and sustain during the recognition process imply flexibility in processing, storing and accessing information. Hence the need for flexible restructuring of memory.

It is because of all the above reasons, with the common denominator of the need for flexibility, that there is the need to study how people do it. It is the flexibility of human processing of sensory information, and interaction with the environment, that is so remarkable and far more advanced than any current machine performance in similar circumstances.

Why do we need such flexibility in the machine's performance? Today's robots are typically reprogrammable machines that can perform a fixed sequence of actions. In the future, however, we hope to have robots which, based on their sensory data and some a priori knowledge about themselves and their world, will be able to adjust their actions depending on the task, the momentary sensory information, environmental changes, etc.

Furthermore, we envision robots in households where, by definition, they will have a much more variable environment than any industrial-factory environment. They also will have to interact in a natural way with people. All this requires robots with much more flexibility than anything available today.

6. References

Bajcsy, R., & Lieberman, L. Texture gradient as a depth cue. *Computer Graphics and Image Processing,* 1976, *5,* 52–67.

Bajcsy, R., & Rosenthal, D. Visual and conceptual focusing revisited. In A. Klinger and S. Tanimoto (Eds.), *Structured computer vision.* New York: Academic Press, 1980.

Baker, H. H. *Depth from edge and intensity based stereo.* Unpublished doctoral dissertation, Computer Science Department, Stanford University, 1981.

Barnard, S., & Fischler, M. A. Computational stereo. *VACM Computing Surveys,* 1982, *14*(4), 553–573.

Bejczy, A. K. Effect of hand-based sensors on manipulator control performance. *Mechanism and Machine Theory,* 1977, *12,* 547–467.

Bejczy, A. K. Smart sensors for smart hands. *AIAA/NASA Conference on Smart Sensors,* Hampton, Virginia, November 14–16, 1978.

Binford, T. Visual perception by computer. *IEEE Conference on Systems and Control,* Miami, December 1971.

Binford, T. O. Sensor systems for manipulation. *Remotely Manned Systems Conference Proceedings,* 1972, 283.

Blum, H. A transformation for extracting new descriptions of shape. In W. Dunn (Ed.), *Models for the perception of speech and visual form..* Cambridge, MA: MIT Press, 1967.

Briot, M. *La stereognoise en robotique application au tri de solides.* Unpublished doctoral dissertation, L'Universite Paul Sabatier, Toulouse, France, 1977.

Clot, J., & Falipou, J. Realisation d'ortheses pneumatiques modulaires, etude d'un detecteur de pressions plantaires. *Publication LAAS,* December 1978, No. 1852, Toulouse.

Clot, J., Rabishong, P., Peruchon, E., & Falipou, J. Principles and applications of the artificial skin. *Fifth International Symposium on External Control of Human Extremities,* Dubrovnik, August, 1975.

Crossley, F. R., Erskine & Umholtz, F. G Design for a three-fingered hand. *Mechanism and Machine Theory,* 1977, *12,* 85–93.

Dane, C. A., III *An object-centered three-dimensional model builder.* Unpublished doctoral dissertation, Computer Science and Information Department, University of Pennsylvania, 1982.

Flatau, C. *Design outline for mini-arms based on manipulator technology.* (Memo No. 300). MIT Artificial Intelligence Laboratory, May 1973.

Foley, J. D., & Van Dam, A. *Fundamentals of interactive computer graphics.* Reading, MA: Addison-Wesley, 1982

Gennery, D. B. Object detection and measurement using stereo vision. *Proceedings of the International Joint Conference on Artificial Intelligence,* 1979, 320–327.

Gibson, J. *The perception of the visual world.* Boston, MA: Houghton Mifflin, 1950.

Grimson, W.E.L. *Computing shape using a theory of human stereo Vision.* Unpublished doctoral dissertation, Department of Mathematics, MIT, 1980.

Hannah, M. *Computer matching of areas in stereo images* (MEMO AIM–239). Stanford University, July 1974.

Hanson, A., & Riseman, E. *The design of semantically directed vision processor* (CIS Department Report 75C–1). University of Massachusetts, February 1975.

Harmon, L. D. *Touch-sensing technology: A review (MSR80-03).* Dearborn, MI: Society of Manufacturing Engineers, 1980.

Harmon, L. D. Automated tactile sensing. *International Journal of Robotics Research,* Summer 1982, *1*(2), 3–32. (a)

Harmon, L. D. Robotic taction for industrial assembly. Working Paper, December 1982. (b)

Hill, J. W., McGovern, D. E., & Sword, A. J. *Study to design and develop remote manipulator system.* Prepared for National Aeronautics and Space Administration, Moffett Field, CA: Ames Research Center, Contract NAS2–7507, 1973.

Hillis, D. W. A high-resolution touch sensor. *International Journal of Robotics Research,* Summer 1982, *1*(2)33–44.

Horn, B. *Shape from shading: A method for obtaining the shape of a smooth opaque object from one view* (Tech. Rep. 79). MIT Laboratory of Computer Science, 1970.

Horn, B. *Sequins and quills—representations for surface topography.* Workshop on the Representation of 3–D Objects, Philadelphia, May 1979.

Inoue, H. *Force feedback in precise assembly tasks* (Artificial Intelligence Memo No. 308). MIT Artificial Intelligence Laboratory, 1974.

Ivancevic, N. S. Stereometric pattern recognition by artificial touch. *Pattern Recognition,* 1974, *6,* 77–83.

Jacobson, S. C. Development of the Utah artificial arm. *IEEE Transactions on Biomedical Engineering, BME-29*(4), 1982.

Jaksic, D. Mechanics of the Belgrade hand. *Proceedings of the Third International Symposium on External Control of Human Extremities,* Dubrovnik, August 1969.

Julesz, B. Visual pattern discrimination. *IRA Transcript on Information Theory* (Vol. II), 1962, *8*(1) 84–92.

Kinoshita, G., Aida, S., & Mori, M. Pattern recognition by an artificial tactile sense. *Second International Joint Conference on Artificial Intelligence,* 1971.

Levine, M., O'Handley, D., & Yagi, G. *Computer determination of depth maps.* Pasadena, CA: Jet Propulsion Laboratory, 1973.

Lieberman, L. *Computer recognition and description of natural scenes.* Unpublished doctoral dissertation, CIS Department, University of Pennsylvania, 1974.

Marr, D., & Poggio, T. A computational theory of human stereo vision. *Proceedings of the Royal Society of London B–204,* 1979, 301–328.

Mishihara, H. K. *Representation of spatial organization of three-dimensional shapes for visual recognition.* Unpublished doctoral dissertation, MIT, 1978.

Mohr, R., & Bajcsy, R. Packing volumes by sphres. *IEEE Transcript On Pattern Analysis and Machine Intelligence,* 1983, *5*(1).

Nerozzi, A., & Vassura, G. Study and experimentation of a multi-finger gripper. *Proceedings of the Tenth ISRI,* Milan 1980.

Nevatia, R. *Structured descriptions of complex curved objects for recognition and visual memory.* Unpublished doctoral dissertation, Computer Science Department, Stanford University, 1974.

Okada, T. Object-handling system for manual industry. *IEEE Transcript on Systems, Man, and Cybernetics, SMC–9*(2), 1979, 79–89.

Okada, T. Computer control of multijointed finger system for precise object-handling. *IEEE Transcript on Systems, Man, and Cybernetics,* SMC–12(3), 1982, 289–299.

Okada, T., & Tsuchiya, S. Object recognition by grasping. *Pattern Recognition,* 1977, *9,* 111–119.

O'Rourke, J., & Badler, N. Decomposition of 3–D objects into spheres. *IEEE Transcript on Pattern Analysis and Machine Intelligence,* July 1979, *1.*

Paul, B. *Kinematics and dynamics of planar machinery.*

Purbrick, J. A. A force transducer employing conductive silicone rubber. Personal communication, 1980.

Railbert, M. H., & Tanner, J. E. Design and implementation of a VLSI tactile sensing computer. *International Journal of Robotics Research,* 1982, *1*(3), 3–18.

Rabishong, P., Stojiljkovic, Z., & Peruchon, E. Automatic Control of the grasp with transducer in the finger. *Proceedings of the Third International Symposium on External Control of Human Extremeties,* Dubrovnik, August 1969.

Roberts, L. *Machine perception of three-dimensional solids* (Techs Rep. No. 315). MIT Lincoln Laboratories, May 1963.

Rock, I. *An introduction to perception.* New York: MacMillan 1975.

Roth, B. *Performance evaluation of manipulators from a kinematic viewpoint.* NBS special publication on Performance Evaluation of Programmable Robots and Manipulators, 1975, 39–61.

Rovetta, A. On specific problems of design of multipurpose mechanical hands in industrial robots, *Proceedings of the Seventh International ISRI,* Tokyo, 1977.

Rovetta, A., & Casarico, G. On prehension of a robot mechanical hand: Theoretical analysis and experimental tests. *Proceedings of the Eighth ISIR,* Stuttgart, 1978.

Salisbury, K. J. *Kinematic and force analysis of articulated hands.* Unpublished Doctoral dissertation, Mechanical Engineering Department, Stanford Univeristy, 1982.

Scheinman, V. D. Design of a computer controlled manipulator (Stanford Artificial Intelligence Project Memo AIM–92). Stanford, CA: Stanford University, June 1969.

Shah, J. *A force-sensitive wrist for a hand-arm manipulator.* 1972.

Shimano, B. *The kinematic design and force control of computer controlled manipulators* (Memo AIM–313). Stanford Artificial Intelligence Laboratory March 1978.

Skinner, F. Design of a multiple prehension manipulator. *Mechanical Engineering,* September 1975, 30–37.

Silver, D. *The little robot system* (Artificial Intelligence Memo No. 273). MIT Artificial Intelligence Laboratory, 1973.

Sloan, K. R., Jr. *World model driven recognition of natural scenes.* Unpublished doctoral dissertation, CIS Department, University of Pennsylvania, June 1977.

Soroka, B. A. *Understanding objects from slices: Extracting generalized cylinder descriptions from serial sections.* Unpublished doctoral dissertation, Computer Information Sciences Department, University of Pennsylvania, March 1979.

Stevens, K. *Surface perception from local analysis of texture and contour.* Unpublished Doctoral dissertation, MIT, 1979.

Sutro, L. L. & Kilmer, W. L. *R–582 assembly of computer to command and control a robot.* Paper presented at the Spring Joint Computer Conference, Boston, 1969.

Sword, A. J., & Hill, J. W. *Control for prosthetic devices with several degrees of freedom.* Paper presented to the 17th Annual Human Factors Society Convention, Washington, D.C., October 1973.

Tanimoto, S. L. *Hierarchical approaches to picture processing.* Unpublished Doctoral dissertation, Department of Electrical Engineering and Computer Science, Princeton University, 1975.

Thompson, W. Disparity analysis of images. *IEEE Transcript On Pattern Analysis and Machine Intelligence,* 1980, *2,* 333–340.

Tomovic, R. The human hand has a feedback system. *International Federation of Automatic Control* (Vol. 2). Butterworth's Scientific Publications. London: 1960.

Tomovic, R. A new model of the Belgrade hand. *Proceedings of the Third International Symposium on External Control of Human Extremeties,* Dubrovnik, August 1969.

Tzikos, S. Personal communication, 1983.

Uhr, L. Layered "recognition cone" networks that preprocess, classify, and describe. *IEEE Transactions, Computers, Vol. 21,* 1982.

Whitney, D. E. Force feedback control of manipulator fine motions. *Journal of Dynamic Systems Measurement Control,* June 1977, 91–97.

Witkin, A. *Shape from contour.* Unpublished Doctoral dissertation, Department of Psychology, MIT, 1980.

Managing the Acquisition of an AI Capability: Some Observations, Suggestions, and Conclusions*

Walter Reitman
ARTIFICIAL INTELLIGENCE DEPARTMENT
BBN LABORATORIES
CAMBRIDGE, MA 02238

1. Observations

During a slow moment last week, my colleague Madeleine Bates stopped by my office. She and Rusty Bobrow, continuing their tracking of current natural language interfaces (see the chapter by Bates and Bobrow), had just returned from a trip to New York, to a firm with such a system. What follows is one interchange they had with it.

> Bates: Are all of the vice presidents males?
> System: Yes.
> Bates: Are any of the vice presidents female?
> System: Yes.
> Bates: Are any of the male vice presidents female?
> System: Yes.

Bates and Bobrow gave the system a total of 70 queries. Of these, 21 were processed correctly; 21 were handled incorrectly (including some accidental right answers); 13 defaulted without retrieving the desired data; and in the last 15 cases, either the system gave up, or the user did.

Perhaps Bates's question about vice presidents is cruel and unusual punishment for a first generation natural language front end. Certainly such results may give an inaccurate impression of the capabilities of other sys-

* I would like to thank Lyn Bates and Rusty Bobrow for making their data available to me. I also want to thank them both for reading and commenting on this chapter.

tems now on or about to come on the market. And given a user who knew this system's limitations, and restricted his questions to sentence forms and subject matter he knew it could handle, quite likely even this system would show a much better hit rate.

Probably so. But for this system at least, even mundane questions sometimes created problems. For example, in response to a query of the form "list all lawyers in Michigan and doctors in New York," the system would respond with a list of all lawyers and doctors living either in Michigan or New York—altogether too much of a good thing.

I draw two morals from this story:

1. You can now type English to a computer, and it can type English back; but unless you know your system well, you'd better not bet your job on what it tells you.
2. We may be watching the beginning of a bull market in artificial intelligence applications (I believe we are). But the early stages of a bull market can be treacherous: buyers beware.

The evidence for the bull market is all around us. The startup activity described in Austin's chapter continues unabated. The Defense Advanced Research Projects Agency hopes to commit $50,000,000 in the coming year to research in strategic computing, growing to a total on the order of half a billion over the next five years. A significant portion of this will fund development of more robust and powerful AI systems.

The transfer of AI technology to industry is a major goal of the DARPA initiative; and as is evident from their enthusiastic response to AI seminars and symposia, business and industry are strongly interested in the prospect. Even now, private industry contracts for multi-million dollar AI systems development projects have been reported, and new AI firms are springing up weekly to meet the projected needs. But remember: The bull market is still very new. Buyer beware.

2. Suggestions

During a break in the NYU Symposium sessions, the Director of Corporate Research for a Greensboro, N.C., firm stopped me in the hall. "Now we've been to an AI symposium," he said. "What do we do next?"

There are several possibilities.

1. Books and articles. The individual chapters of a book like this one contain a great deal of information on the strengths and weak-

nesses of expert systems, natural language interfaces, and other AI technologies. The experts don't yet agree on a single picture, and many of them have a stake in one approach or another. But if you approach the problem of figuring out the prospects for AI much as you'd shop for a good used car, you'll soon get an overall feeling for the issues and arguments.

2. More seminars. Several organizations offer more intensive experience and training in AI. This costs more than (1), but there's a limit to what you can understand without some hands on experience.

3. Consultants. A number of firms with established consulting practices are expanding to provide services in the AI area. Such organizations do not, in general, supply specialized AI products or services of their own. Thus, if you can find one with demonstrated AI competence, you should be able to obtain a disinterested appraisal of your AI needs, and of the relevant options available to you.

4. AI product, service, and development firms. With the explosion of interest in AI applications, a variety of suppliers are moving to meet the demand for AI consulting sevices. Some of them are single product firms. Others specialize in AI applications for a particular industry group, e.g., finance, insurance, and banking. Still others are established organizations with major investments in a broad range of AI research, applications, and development capabilities. In general, all such organizations are likely to view consulting primarily as a vehicle for marketing their products or system building services. But since they are in the AI business, the best of them will bring to their consulting a level of competence and experience not easily matched by conventional consulting firms. And the broad-based firms, having a range of AI technologies to draw upon, are likely to be able to provide services well matched to the specific requirements of your firm.

5. Demonstrations. Vendors of AI systems, products, and tools will be glad to show you what they have to offer, and most of them will conduct such demonstrations without charge. But making a business decision about AI mainly on the basis of such demonstrations could prove expensive in the long run. As the little story at the beginning of this chapter suggests, you probably do well to defer this step until you have either acquired some AI sophistication, or located a consultant you have confidence in.

6. In house projects. Assuming you have a good computing or MIS staff, setting up a tutorial project in house is a good way to gain experience. Working with an organization with AI experience, or perhaps with a local university, you can achieve a significant amount of technology transfer at relatively low cost. DEC's AI experience

(see the chapters by McDermott and O'Connor) began along these lines.

7. In house AI applications staff. If the applications you are interested in do not require development of new AI technology, gradually improving the expertise of your existing computing or MIS staff may give you an adequate AI applications competence. DEC's experience is an outstanding example of this strategy. A similar approach is being taken abroad. In West Germany, for example, the government will provide subsidies to industry covering up to half the salary of a presently employed staff member who undertakes to acquire AI competence.

8. In-house basic AI capability. If you have big plans for AI in your organization, if you want to develop new AI technologies to enhance your existing business, if you want to promote a high level of interaction between your AI experts and other technical groups in your firm, and if you want to keep the results totally under your control, this is the way to go. It also is the most difficult, time consuming, and expensive strategy to follow: A number of firms are trying it, but few appear to be succeeding. Among the difficulties are these.

a. Cost. Direct costs for labor and capital for a group of 30 people could easily run to $5,000,000 per year.

b. Availability. There are far more available jobs at the moment than there are people. What will you offer good people that they can't get elsewhere?

c. Management style. With all the job opportunities available, good AI professionals can afford to pick and choose their environments. If they don't have fast, easy access to a management that appreciates and understands their interests as they see them, if they feel that management decisions come too slowly or that management seems unresponsive to technical problems and opportunities, if management lacks a real concern for their well being as individuals, if they don't enjoy working with their colleagues, or if their salaries are not competitive (regardless of what this does to your personnel compensation plans), they typically have no trouble moving elsewhere. If you are trying to build a good basic AI research group, you will need to develop sufficient esprit and satisfaction to either offset your not being in Los Angeles, Cambridge, or the Bay area, or to keep people from making a local phone call and taking a new job across the street if you are.

What if you try to reduce the costs and simplify the problem by shooting for a limited, very specialized AI group? The strategy may be feasible, but it has its own problems. Suppose you focus upon natural language understanding, for example. You need at least three senior people, and four or five junior staff, or else the group will feel lonely and isolated. But if you are going to do natural language understanding right, you need some knowledge representation people. And as the papers by Webber and Woods suggest, you very likely also will need some people who know about users' intentions and plans. You will need still more people if you want to incorporate your natural language capabilities in expert systems, and you certainly are going to need some AI tools and environments people (see Sheil's paper) if you expect your group to work really effectively. And so it goes.

It is not that the strategy of building an excellent in house group of basic AI researchers can't work. Though the limited supply of good AI researchers, and the intense demand for them from universities, research groups, and business and industry, exacerbate the difficulties, the basic problems are fundamentally no different from those to be encountered in building quality staff in other technical areas. With sufficient support, motivation, and managerial attention, the strategy certainly can work. But without these factors, and without an adequate understanding of the problems involved, the chances of success are small.

3. Conclusions

If you are looking into AI possibilities for your firm, you face two problems:

1. Figuring out what AI is; and
2. Figuring out what your organization should do with it.

Most of this book has been about the first problem, so let's focus here on the second.

What are your interests in AI? Do you want to add a little AI pizzazz to your current product line to give you an extra marketing edge? Do you want to improve your current computing and MIS systems? Do you want to market AI products or services? If so, what kinds? And if so, do you want to license them from AI houses, or design them yourself? Or, most ambitious of all, do you want to use AI as a basic engine, driving your business to new developments on many fronts?

The best way to deal with these two problems is in parallel, using what cognitive psychologists call progressive deepening. You find out a little about a lot, and then as you get some first ideas about the specific areas you really need to know more about, you increasingly concentrate your attention on those specific options.

Different firms will have different answers. So, too, with the primary AI suppliers themselves. The result, I expect, will be a highly differentiated, multi-tiered AI industry. There will be users, retailers, wholesalers, single-product firms, specialists in one or another application area, and a small number of broad-based research and development organizations. All of them will have their own ideas about the way things are going, and their own objectives when you come to deal with them.

In making your way through this maze of AI opinions and options, you will need persistence and, above all, patience. You shouldn't expect neat, simple answers, or even consistent ones. But with persistence, patience, and an adequate supply of salt, you should be able to figure out what is going on, and what you can do with it.

Just watch out for programs with odd ideas about your vice presidents.

Author Index

Pages in *italics* indicate where a complete reference can be found.

Subject Index